RUSSIAN CHALLENGES

INTERDISCIPLINARY STUDIES
ON CENTRAL AND EASTERN EUROPE

Vol. 8

Edited by
Christian Giordano, Nicolas Hayoz & Jens Herlth

PETER LANG
Bern • Berlin • Bruxelles • Frankfurt am Main • New York • Oxford • Wien

GALINA MICHALEVA &
ANDREY RYABOV (EDS)

RUSSIAN CHALLENGES

Between Freedom and Energy

PETER LANG
Bern • Berlin • Bruxelles • Frankfurt am Main • New York • Oxford • Wien

Bibliographic information published by Die Deutsche Bibliothek
Die Deutsche Bibliothek lists this publication in the Deutsche Nationalbibliografie; detailed bibliographic data is available on the Internet at ‹http://dnb.ddb.de›.

A catalogue record for this book is available from *The British Library,* Great Britain.

Library of Congress Cataloging-in-Publication Data

Russian challenges : between freedom and energy / edited by Galina Michaleva and Andrey Ryabov.
p. cm. – (Interdisciplinary studies on Central and Eastern Europe, ISSN 1661-1349 ; v. 8)
ISBN 978-3-0343-0529-7
1. Russia (Federation)–Politics and government–1991- 2. Petroleum industry and trade–Russia (Federation) 3. Gas industry–Russia (Federation) I. Michaleva, Galina. II. Riabov, Andrei.
JN6695.R865 2010
338.2'7280947–dc22

2010046979

FNSNF
Swiss National Science Foundation

— GEBERT RÜF STIFTUNG —
WISSENSCHAFT.BEWEGEN

Cover illustration: from iStockphoto.com © by Csaba Peterdi

ISSN 1661-1349
ISBN 978-3-0343-0529-7

Hochfeldstrasse 32, CH-3012 Bern, Switzerland
info@peterlang.com, www.peterlang.com, www.peterlang.net

Printed in Switzerland

Table of Contents

Editorial

"Russian Challenges: Between Freedom and Energy", June 23 2006. That was the name of the conference at the University of Fribourg in Switzerland in the summer of 2006, organized by Nicolas Hayoz, professor at the University. The speeches of that conference became the foundation of a collective treatise that is presented here. It seems at first that a lot of changes have occurred in Russia and the world since this conference took place, including the international financial crisis, the drop in energy prices, and the election of a new president in Russia. But the presented texts have not lost their relevance. The economic structure and political system of Russia have not changed.

The main idea of the conference, as well as that of the collective works, was to show that the energy-exporting nature of the economic system during an oil and gas boom resulted in the formation of a vertically integrated political system that is characterized by high stability. Its most distinct trait is a revenue-based economy. Authoritarian tendencies are increasing in politics; the role of independent actors in relation to the government is gradually declining while the elites in power continue to present their corporate interests as being national ones.

In relation to this, most of the articles presented here analyze the influence of the factor of oil and gas on various sides of internal and foreign policy.

Russia's economy still remains absolutely dependent on the export of energy resources, primarily oil and gas. In addition to that, the global crisis, which has had a deep impact in Russia, clearly demonstrated the vulnerability of Russia's export-based resource economy and its slow adaptation to new challenges. The stated projects of modernization and innovation remain nothing more than declarations of intentions. The current elites, who are objectively interested to retaining and strengthening the country's existing economic system, are not ready to initiate great changes even though the crisis affirms the necessity of such changes. Also, since the possibility for independent political and civil action is limited, the economical and social systems of the country have become difficult to reform. Some of the articles in this collection examine such ideas.

The specifics of the political system are the same today as at the time of the conference. If anything, the authoritarian traits of the regime have intensified even though the head of state has changed: political decisions depend even more on the conjuncture and the positions of people at the top of the pyramid of power, elections have lost their competitive nature, the number of political parties has declined, the pressure on NGO's has increased, and the mass media, especially television, have become a propaganda tool. The main specifics of the political structure in Russia are the domination of informal institutions in decision-making and the bonding of government and business. In the struggle between competing businesses - political groups of their own - corruption on every level of government has intensified.

Russia's foreign policy has retained its multilateral approach, but energy resources are being used as instruments of political pressure more frequently on partners. Russia openly began to voice her claims to dominate in the post-Soviet region, moving from imperial rhetoric to the use of force in the conflict with Georgia in the summer of 2008. These events forced the European Union to concentrate their attention on Russia regarding matters of European security and the stability of energy shipments on the one hand, and seek new for forms of interaction with other post-Soviet republics, thus avoiding the Russian Federation, on the other.

The aftermath of the financial-economic crisis increased the socio-economical gap between the different kinds of regions and forced specialists to ask whether the Russian Federation will continue to exist within its current borders if ongoing trends continue.

All of these problems are presented in the collective treatise in different degrees of acuteness. In addition to the participants of the conference, the authors include leading Russian specialists in energy development and transformation of Russia's political institutions. The authors employ varying theoretical approaches and have different ideological views; they present contrasting views of the problems and processes discussed in the treatise.

In the introductory article by the famous Russian politician and chairman of the social-liberal opposition party "Yabloko", Grigory Yavlinsky, Russian economic perspectives of that time are characterized as the "forecast" of the crisis, and the dependence of the Russian economy on prices for energy resources is stated. Also emphasized is the fact that the ruling elite was not ready, despite the comfortable economic conjuncture, to carry out reforms that would modernize the country.

The article of political analyst Andrey Ryabov shows how the policy of the "petrol state" does not create conditions for successful modernization, but instead crowds social elevators, conserving the marginal status of some social groups and therefore the status quo.

Vladimir Milov and Nina Pussenkova, both well-known specialists in the field of energy and energy policy, analyze the role of oil and gas on the Russian economy in great detail based on various economic indicators, as well as the state and perspectives of their production in Eastern Siberia and the Far East.

An expert in the field of international relations, Fyodor Lukyanov, analyzes the key role of the export of oil and gas in Russia's foreign policy and the possibility they offer in developing relations between Russia and the European Union.

The leading Russian ecologist Alexej Yablokov examines the parameters of a process that he calls the de-ecologization of Russia.

Regional differences and the specifics of development in various regions of the Russian Federation have a decisive meaning for the country's future. The article of Natalia Zubarevich, a leading specialist in regionalism, analyzes this problem. She examines the polarization of Russia's regions in terms of presence of resources and paces of development.

Some articles are dedicated to the specifics of contemporary political development in Russia. In this way, the historian Dmitrij Furman analyzes the possibility of new historic cycles and alternatives in Russia's development in the context of historic traditions and the country's political experience. The phenomena of the current regime's political stability and the peculiarities of interaction between government and the people are the themes of political scientist Maria Gordeyeva's article. Political scientist Alexei Zudin gives an answer to the question whether the Russian monocentric political model has the ability to evolve and discusses institutional changes to the existing system.

Galina Michaleva, a representative from the management of the opposition political party "Yabloko" writes about the conditions and possible actions of oppositional political forces in the reality of an authoritarian regime in Russia.

The Afterword, being based on the positions of the treatise authors, takes a quick look at the events that occurred from the publication of the articles up until the middle of 2009.

Knowing that in Switzerland, in Europe, in the West, and in the world as a whole there are a lot of specialists who understand the specifics of Russia's economic, social, and political development, following the changing situation very carefully and giving their in-depth reviews, all authors of this collection of articles are Russian. "Russian Challenges: Between Freedom and Energy" is thus an examination of the events of one country by leading experts from that country. We hope it will be of interest to English-speaking readers.

Galina Michaleva and Andrey Ryabov

Grigory Yavlinsky

Anticipating the Crisis: Russian Economic Prospects

When estimating Russia's current economic prospects, the majority of today's economists are brimming with optimism. Over the next several years their predictions include GDP growth rates of seven to eight percent, a continuing consumer boom, growing industrial investment and economic diversification, all combined with a further decrease in inflation and a stable public financial system. Even considerable drops in oil prices and commodities production are not seen as substantial detriments. Despite a swift decline in Russia's trade surplus and currency reserves, the rouble will maintain its long-term stability and grow stronger against the majority of the world's currencies. All of this, happening under politically stable conditions, will further improve Russia's investment climate, promoting foreign capital inflow and thus supporting high rates of economic growth. Furthermore, in this dream world, the tendency of Russian big companies towards globalisation will work to the advantage of the Russian economy, offering ever more financial flows and administrative resources.

Impressive, but Only Mildly …

These promises of a rapid economic miracle are based on current achievements. But Russian economic success, though indisputable, is not as great as it might seem and, in our opinion, does not warrant such optimism.

Firstly, we should not be deluded by the high GDP growth rate. Statistics by nature can be deceptive, and when applied to something as lacking in diversity and transparency as the current Russian economy, that deceptive nature increases exponentially. We do run into developing economies that annually report improbably high GDP growth rates of twenty percent or more. However, for some reason, this growth in these countries is never felt by the general population, nor is it felt by the private investors who should be lining up for a chance to invest their money. Of course, Russian growth rates tend to more accurately reflect the real situation, but under current pre-election conditions, with a very high occurrence of unregistered business activity and a high rate of inflation, our system of monitoring and assessing the scale of economic activity seems to show, at best, a very rough picture.

The official statistics from this year (2005) are surprising. For example, industrial investments rose by twenty percent during the first few months of this year compared with the previous year after adjusting for inflation. May and April saw a sudden skyrocketing in prices for manufactured industrial goods - an indication generally considered as more reliable and accurate than other measurements. Suddenly, industrial production growth rates, which had shown low volatility and a general downward tendency over the last several years, started to rise inexplicably, especially in the manufacturing sector. These drastic changes in key economic indicators, without any glaringly obvious changes in the general conditions of economic activity, suggest that our statistical methods remain only minimally useful for recognizing important changes in development and real economic processes.

Secondly, the figures on foreign investment, which optimists use as proof of further growth prospects, tell us absolutely nothing. To a large degree, these "investments" reflect the movement of assets between various branches and affiliates owned by specific individuals whose businesses have Russian roots. For reasons that have little to do with investment in real economic sectors, these assets constantly cross into and out of the country. As a result, the scope of the "accumulated" foreign direct investment (minus the outflow of capital investment) in the Russian economy is, to put it mildly, less than overwhelming. The statistics on capital inflow into the private sector, as defined by the balance of payments, are also not indicative, as they mainly reflect either the attraction of assets by state or semi-state corporations, or the movement of short-term speculative capital.

Thirdly, stability and growth in the financial sector, passed off as one more example of successful economic growth, should be considered with caution. These rates are predominately derived from a combination of very low starting levels and an abnormal boom of consumption and short-term financial investments. As for the rise in manufacturing, this illusion is generated by government regulation and the drawing in of latent financial streams to the industry. Those familiar with the sharp rise in manufacturing associated with periods of "catch-up development" in other economies understand that what is occurring in Russia today bears no resemblance to the processes of accelerated industrialization that has already changed the economic landscape of many countries.

Recession? No Question!

If the issue here was only the need to reassess the current virtues of the Russian economy, it could fairly easily be ignored. The real problem, though, lies in the unrealistically high expectations of continuing, guaranteed growth over the medium and long term.

Here, I must momentarily digress. It is obvious that the cycles of economic dynamics (growth vs. correction or recession) possess certain inertia. This means that economic phase changes occur with some delay. Economic growth always begins some time after an objective basis for growth has been established. It logically follows that even after the impetus for such growth has subsided, effects of growth can be seen for a certain (sometimes lengthy) period of time. Naturally, the more vigorous and enduring a period of growth is, the more vigorous the inertial forces behind it become. This allows even "baseless" growth to continue, and can even permit growth in spite of opposite objective circumstances. However, in truth, this only results in a more dramatic and painful collapse when economic dynamics are brought down to conformity with actual economic conditions.

One more defining factor in the scope of the "delay" during the changing of cyclic phases is the general level of economic development of each individual national economy. The higher this level is (with all the other factors being identical), the less volatile (an indicator of variability) the markets are. Thanks to their complexity, developed and diversified economies are more difficult to "accelerate" (thus for developed countries a 2–3% GDP growth rate is considered sufficient, while a rate of 5% is seen as a cause for concern), while at the same time macroeconomic downturns tend to be far less dramatic. Decline in these economies is more likely to come in the form of specific stagnation set against a background of high levels of consumption and reasonably optimistic expectations.

Less developed economies, on the other hand, can experience periods of rapid manufacturing growth and consumption. However, such periods almost always lead to an unjustified increase in expenses, reduced efficiency and overestimated expectations. The consequence is unavoidably a financial crisis, caused by the need to resolve the contradiction between the unjustifiably high expenses of producers and their limited productive basis.

Capitalism in Russia is not the capitalism of the G7 countries. These countries long ago endured the the sicknesses from which Russia suffers today. Russian capitalism is underdeveloped (or "developing," to use a more politically correct expression) and high volatility and general instability are the norm. The latter is connected, first and foremost, to the absence of mature social institutions and to the psychological vulnerability of the entrepreneurial class. This vulnerability comes from uncertainty in the future and concern over the respect of property rights by the state and society. In such an economy, sudden transitions from high growth rates to plummeting production are hardly anomalous, and, in fact, are likely the natural characteristic of economic dynamics. The question is therefore not whether current economic growth, already in its eighth consecutive year, will invariably lead to recession. The question is only when, and on what scale.

Growth Exhausted

At the moment the objective basis for economic growth is completely exhausted. Any further growth will require, at the very least, a temporary pause while a new set of objective preconditions are formulated. A serious cooling-off of the market is not only probable, it is inevitable; from the standpoint of long-term growth conditions it is even desirable. Unrealistic expectations on the part of some important economic actors, mainly the state and the companies under its control, are leading to a schism between the movements of basic macroeconomic indicators and the dynamics of deep, fundamental forces. This schism continues to widen, ensuring the inevitable correction will be even more dramatic.

There are at least a few objective reasons that support these statements. It is clear that the factors that led to growth and upward economic trends in Russia in the opening years of the 21st century have already lost their force and influence. The first of these factors was the low cost of production, including the prices of energy, labour and a considerable number of business services. Secondly, there was the effective protection of the domestic market caused by an undervalued rouble and the sharp increase of financial opportunities for Russian companies resulting from the unexpected rise of world prices for raw materials and the ease of access to extra financial resources. Lastly, there was the positive effect from several improvements in state institutions: the normalisation of accounting practices, the end of salary withholdings for state employees, the reform of government financing regulations, the reprioritizing of the direction of public funds, and other similar measures.

Today, for all these indices, the situation has either returned to where it started, or completely lost its advantageous effect. By now the comparative advantage of domestic manufacturers through the exchange rate policy has essentially been reduced to nothing. Economically significant import tariffs have been retained for only a very limited range of goods, and there are no other obstacles to prevent the aggressive replacement of domestic production through imported goods. We have seen this clearly over the past several years in the increasing rate of imports, which has greatly exceeded the growth rate of domestic production.

Attempts to contain rising costs have been ineffective. The rise in prices for exported raw materials has generated a stream of "easy" money which has been exploited by everyone who exports any goods or services, including those of a corrupt or criminal nature. The lack of willingness (or ability) on the part of the government to rein in the expenses of state-run monopolies, or to reform the appropriate structures in order to create mechanisms of genuine competition within them, certainly did nothing to stabilize prices. The better part of these monopolies remain to this day just as lacking in transparency and resistant to governmental control as they were eight years ago. As a result, the overwhelming majority of today's manufacturers face rising costs across the board. The prices of raw materials, energy, labour, transport, and both official and unofficial forms of extortion from authorities are increasing by double-digit percentages in terms of both rouble and foreign currency value. The growth of industrial manufacturing over the last two to three years, limited as it is, can largely be explained by the introduction of new manufacturing capabilities. However, the decisions to invest in these new capacities were made at the very beginning of the decade, back when nobody expected such a long period of high oil prices or such a daunting rise in costs.

As part of the efforts to strengthen the state, Russia is on the exact opposite course from that at the beginning of the 21st century. The size of the bureaucracy and the share of the economy it controls are both rapidly increasing, while the efficiency of the state to fulfil its functions in the economic sphere is falling. To see this, one only has to take the time to examine the medium- and long-term projections of the government in the last four or five years and compare them to reality. Additionally, decreasing the scale of corruption turned out to be impossible and was given up.

The results of this are not exactly broadcast, but they are to be expected: rising profits are used to finance a rise in imports and an outflow of capital abroad, while the whole process is termed to be the globalisation of Russian business. Long-term investment in industry is as lacking as it ever was, and the consumer market is up for importers to buy out. With the exception of alcohol and certain building materials, the consumer market share for domestic producers is decreasing in practically all commodity groups.

Multiplying by Zero

In sectors with state participation even stranger and more dubious processes exist. Owing to loans of unprecedented size, large state companies are buying up the activities of the private sector and actively participating in financial investments – a practice of debatable economic expediency from the point of view of state interest. The control over financial streams in this sector is either lost, or, more often, built into the logic of a functioning parallel economy. The state sector is then quickly filled with leaders from areas that have nothing to do with the actual economic activity.

Simultaneously, proprietors in the private sector have become increasingly anxious. They constantly move to export the centre of their entrepreneurial activity abroad while avoiding any publicity or demonstrations of ambition within Russia itself. Illustrations of this process can be seen in any serious mass media outlet, and, for those who professionally observe Russian business life, it has already been obvious for at least two years and is in no way connected to the upcoming elections.

Under such conditions, continuing high rates of economic growth over a period of several years are probable only under the influence of a great force of previously accumulated inertia, and on the condition of enduring high prices for energy carriers. True, at the moment the latter looks quite probable in both short- and intermediate-term prospects. Nevertheless, high world prices for raw materials in and of themselves are not enough to support years of growth if the objective grounds for growth have already washed away. It is impossible to tell in advance when exactly moods and tendencies will change, or what exactly will serve as the trigger. It is just as possible that this will occur a year and a half to two years from now as that it will occur in a few months.

Unlike in developed economies, the phase change is not likely to be initiated by a price drop of the stock market.[1] The stock market today plays only a secondary role in the Russian economy, and its situation is not critically important for the economic behaviour of either entrepreneurs or individual households. By and large the financial sector does not strongly influence the behaviour of the Russian business elite, so the prediction that sharp deterioration there will be the immediate cause of a general economic crisis is hardly justified. It is possible that the reason will be a sharp strengthening of the political (or more precisely, non-economic) risks, which are clearly underestimated. These risks are best understood as the degree of predictability of election results and the make-up of government agencies. Of course, Russia is not Western Europe, where all of this does not generally influence the activity of companies. But in Russia, big business has the flexibility and the resources to find a way into any of the government agencies, whose behaviour and composition can vary sharply from year to year.

From an economic standpoint, the main risk present in an authoritative, vertical-management system is a loss of direction – the middle and bottom sections begin to feel so invulnerable that they bring to naught supervising decisions made at higher levels, taking these decisions and effectively "multiplying by zero". They exclusively act according to personal interests and plans, thereby depriving the system of internal order and logic, even if that logic had been faulty. It is basically a form of "bureaucratic lawlessness", though in practice so-called "power structures" play the most important role here; without them this lawlessness can neither arise, nor exist. This kind of risk is undoubtedly present today in the Russian economy, and is growing tangibly from month to month.

Serious doubts exist concerning the ability of Russian authorities to appraise the situation correctly, to moderate their official

1 Although the current (beginning in the fall of 2008) financial crisis in Russian economy manifests itself mostly in the form of free-falling stocks and dramatic shortage of available financial outsourcing, we believe that this is only the tip of the iceberg. The deeper problems of the Russian economy are institutional in nature. This also explains why the current economic decline in Russia is much sharper than almost anywhere else (author's note).

optimism, and to take measures to prevent an uncontrollable bureaucratic engine from transforming the inevitable cooling of the market into a real full-scale crisis in the near future.

* * *

In addition, Russia's current political situation and its prospects bear mentioning.

The current political system could hardly have developed if Russia did not have its supply of "addictive" resources. Over the last seven to eight years, these have completely corrupted the Russian leadership and elites. High and rising prices for raw materials have become the economic basis for a political system where there is no independent court or parliament and no politically free mass media on a national scale. The executive holds a defining influence over the outcome of elections, and there is a complete lack of civilian control over intelligence and law enforcement agencies.

In truth, under the leadership of Vladimir Putin, Russia has seen the rebirth of the one-party system. As invariably happens, this system has created a much more primitive economy for Russia, leading to inflation and a total violation of private property rights. The combination of political monopoly with private economic interest is the basis of an inefficient and dangerous economic system which deprives three quarters of the able bodied population of any long-term economic prospects.

In general, the whole Russian administrative-authoritative system is extremely unstable and, despite high incomes from energy resources, essentially always contains high political risks. The system runs not on the basis of law, but with the help of law. Various governmental problems are solved by selective application of the law, cases are fabricated in courts, and in some cases defendants are found guilty and sentenced in politically motivated trials. Dissidence is suppressed with the help of media and ideological manipulation.

Over time a new ideology has formed: "We are sovereign – we do what we want. Nobody has authority over us!" The Putin government not only justifies the system of state-sponsored terror of the 1930s that killed millions of people, but it presents this system as being rational. It is therefore hardly surprising to see the criminal use of physical violence to intimidate and suppress all alternative political activity.

Under these conditions, we are naturally left to wonder how we can rationally talk about the European integration of Russia. The basis of Putin's foreign policy is the so-called "realism", or more simply, the "trade and purchase" approach: gas in exchange for advanced European technologies, access to resource deposits in exchange for a share of Western pipelines, positions in boards of directors in exchange for public support, etc. Some Western partners of Russia have accepted this situation. In their eyes it is impossible to get any more from Russia; all available chances of building of real democracy in Russia are gone and they must therefore deal with what is. Such a position is a reflection of the confusion in Western politicians before Russian authoritarianism, who see before them no clear tools to influence the situation.

We stand with those who believe that no matter the complexities, no matter how long and twisting the path may be, Russia can and must become an organic part of the Western world – that a full-fledged European integration of Russia is not simply the best, but the only strategically viable way of development. For Russia, the path away from the European world will quickly lead to a dead end. To start down the right path, our starting point must be our distinctly Russian cultural and historical sense of self-identification. We must answer once and for all the important questions: To what civilisation does Russia belong? What relationship do we have to the idea of communism, which dominated for 70 years? Or with the idea of an empire, even more historically implanted in our history? How do we see our future?

Our own political elite refuse to answer these questions, trying instead to parade before us one idealised caricature after another: the czarist Russia "that we lost", the "order" of Stalin's totalitarian empire, the "stability" of Brezhnev's 1970s. This ideological collage is then combined with aspirations of European resort vacations.

In our opinion, the realization of a European, democratic way looks far more natural to our country. This alone must be the basis of every serious Russian long-term strategy.

Andrey Ryabov

Russian Reforms: The Path of Modernization or Optimization of the Existing System

Reforms in Russia have been objects of heated political discussion throughout our country's post-communist history. The main argument of advocates of reform is usually that no alternatives exist to a certain course. Reforms have been initiated at various times by the Russian government for the last 15 years. Critics of reform claim that in all this time Russia has not moved a step in the direction of a structural economic reform but that the dependency on oil and gas exports has increased. They point out that one of the failures of reform, affecting the country's development to this day, is the huge economic gap that divides society. The division is between the new elite and the general population; this in turn produces tendencies to hold back social and political dynamics. Opponents of the reforms have advocated other, alternative reforms on more than one occasion, but their suggestions have not been implemented.

In the current situation, when the Russian government has stated that its goal is to realize the next phase of market adjustments, the old argument about the substance of reforms has once again become relevant. Also, the first such experience – the monetization of aid services (for certain groups of citizens) – almost ended up as a serious political debacle. Nevertheless, despite the similarities of today's reforms and those of the 1990s, some key differences remain. Understanding these differences will help us better understand Russia's modern development and to define to which degree the current changes fit the country's goals of modernization. This article will focus on these problems.

One of the main problems of Russian reforms that always results in enormous drawbacks is that the originators of reform sacredly believe in the technology of their realization. This derives from the stable belief that Russia remains a very conservative country where most people do not understand and do not realize the importance of reforms. Thus the only chance to make significant changes is when a popular and trusted government exists that is able to execute tough and decisive actions with the support of the people. It is easy to see that an old tradition of Russian political culture lies at that the root of this belief: all changes are made by the critically thinking minority. The socially inert masses are left with the choice of obedient compliance.

Nevertheless, the system of reforms that were mostly dependent on a strong national leader had a solid foundation during the final period of the Soviet system. Old institutions and social relations were falling apart; they were not immediately replaced by new ones. Most people had only a vague knowledge of the ongoing reforms and did not really know what the result would be. Under these conditions, the president was the recipient of an immense social trust: he was not only the cornerstone of social stability and the guarantee against social disintegration, but also the leading force behind the reforms. Boris Yeltsin felt all the specifics during that period of time when he appeared like a biblical prophet and proclaimed on TV: "I ask you to support your president". Reading in between the lines that meant: "Believe me, follow me, and support what I am doing". Society easily agreed to such conditions because the end of the communist era was a revolution. Any revolution, as history shows, is accompanied by a large outburst of utopian thinking. The most unrealistic goals are perceived as easy errands. It is the feeling that imminent happiness is just around the corner that makes people cope with great social faults and deprive themselves of many things. This is what happened in 1992 when the government of Egor Gaidar was pursuing a policy of "shock therapy".

However, in 2005 reforms were being passed in an entirely new country. It was the time of post-revolutionary stability. The mass consciousness had become more rational inside the country, and – from the top to the bottom of the social ladder – social relations were formed. Under these conditions, society in general could not remain a passive player that accepted only the will of the political classes. It now had a chance to become the partner of those who were in power and expected that its interests would at the least be taken into consideration when a new political course was formulated.

However, the old methods of administration were still applied, and the indicated course was carried out around the dominant figure of the president who had popular support. Additionally, compared with the 1990s, current conditions were better for reforms: at that time the communist opposition interfered with Yeltsin's reforms but now there was no one standing in the president's way, which enabled him to follow through with the policy that he saw fit. Analytical mistakes led to protests of pensioners and other groups whose interests were at stake with the monetization of free civil services. The government was forced to make corrections, which watered down the original version of the plan.

A logical question arises from all of this: why can't the governing circles of government stop using this model of reforms, despite its ineffectiveness? Clearly, the reason is not due to intellectual mistakes or getting carried away but to established interests. The fact remains that radical market reforms, as international experience shows, can be an effective instrument only under two conditions: transparency of the political and economic systems as well as public control over the government. Otherwise the reforms become a mechanism of property redistribution in the interests of those who are in power and are able to use administrative resources for personal interests or for the interests of a group. The result is the growth of the gap in society which divides the ruling minority and the rest of the people, and raises political tension in society as a side affect. Evident proof of this is the experience of a number of Arab countries which created statist systems (Tunis, Algeria, Egypt). The liberalization of these systems through market reforms while keeping autocratic political practices with no transparency led to a new social polarization which stimulated new socio-political conflicts and fundamentalist movements. As a result reforms had to be halted in Tunis in order to maintain social stability; a lengthy civil war erupted in Algeria, and there is the possibility of one in Egypt after its president, Hosni Mubarak, leaves politics. During his reign he contained protest sentiment using heavy-handed authoritarian methods. Of course, comparing Russia with Arab regimes may offend our worldview. Nonetheless, it does not change the fact. Concerning the process of changes, the Russian type does not differ from other countries. Ongoing market reforms during the last 15 years of our country's history occurred in a non-transparent environment with no control of the political power, and produced the elite's conservative mentality as well as social gaps that are common in third world countries but not in developed and industrialized nations.

From this experience we can come to a conclusion of what political technology should be implemented in Russia when devising a reform policy. When I use the term "political technology" I don't mean the process of achieving change by using manipulation (usually median), but the technology of change itself. This technology should come from the principle that decisions about the execution of a reform can only be made after a long discussion between government institutions on one side and civic organizations – including, of course, political parties of the opposition – that represent social groups whose interests will be affected by the reform. The statement that the Russian government usually uses – that such groups in Russia are so weak and disorganized that they cannot be reliable partners in the negotiation process – is false. Such organizations do exist and it is important for the government to have the desire for broad dialogue and to seek institutional mechanisms for the decision-making process. It is evident that, if the decision is agreed upon by all sides and its acceptance was preceded by an open public debate to determine reform borders upon which the society can agree, such an approach will lower socio-political drawbacks.

Difficult reforms should be passed gradually, not as a single packet, as was done in the case of monetization and the administrative reform. Monetization required the maximum concentration of the state apparatus, but that apparatus was more concerned with its own reorganization. The outcome was very unprofessional. We also have to take into account that a society's potential for adjustment is limited. It has its own boundaries depending on the people's quality of life, their social experience, and the socio-cultural specifics of the country and its regions.

In any case we need to clearly see that the time of "vertical" reform technologies – when several people in government think that only they know what to do and how to do it and society just has to obey – is in the past. Such reforms will not have any positive affects, neither today nor tomorrow. Only by realizing reforms through dialogue and discussion methods can we break the repeating cycles of Russia's political system. After each radical reform, politics of social pacifying occur that are directed towards raising the status of certain social groups or of the people as a whole; this usually happens shortly before national elections. It is important to point out that such a policy does not change the nature of social relations but only conserves social paternalism in its traditional form.

The suggestion made here brings us face to face with the main problem: Are reforms in Russia an instrument of modernization for the existing social system or do they have a different meaning? In order to figure this out we have to start using the term "petrol state" in our analysis. In the last year this English term came into use; it refers to Russia's modern economy and, to a smaller extent, to its social system. In its broadest definition, the term petrol state implies the critical dependency of the economy on oil production and export.[1] Even though this definition is nominal in the case of Russia (for example, because gas is also a strategic export resource), it clearly expresses an objective that is of interest to us. The Russian Federation has tightly integrated itself in the global system as a supplier of fossil fuels, primarily energy resources for industrialized and industrializing nations. The stability of Russia's internal order and her place in international relations depends upon fulfilling this role.

The petrol state is not only the above-mentioned economic model and government mechanism which guards and reinforces it, but also an organization with a structure of social relations and a hierarchy of social groups. According to the findings of economists, the modern Russian state, whose economy is based on energy export, can provide a "west European" standard of life for no more than 50 million of its citizens.[2] The social groups that benefit from this system are represented not only by those who are related with producing refining and transporting energy resources. It is also the people who work in related sectors, such the advanced legal-financial sector that services the oil and gas routes and controls that the cash flows to certain banks. These are the people who work in a well-built propaganda machine, whose task it is to convince their fellow

1 Нефтяное государство // Новая газета, 2005 г., 96, р. 10.

2 См. Явлинский Г. Демодернизация. Современная Россия: экономические оценки и политические выводы. М., 2003, pp. 55–56.

citizens that the Russian petrol state is just a good as any developed democracy. It is also the large bureaucracy that plays a key role in the distribution ("sawing off", using modern slang) of money of the ever-growing budget pie. And it includes the workers of other exporting industries, usually not the leaders of the 21st century in terms of science and technology (metallurgy, general chemistry). In other words, the beneficiaries are those who are included in some way in the consumption of resources coming from export. The other, more than 90 million, who do not have such access (primarily large groups of people with government salaries and pensioners) are forced not only to live in different economic conditions, but also in a different social reality. This concerns the inhabitants of large and medium-sized cities. They not only have far lower standards of consumption but also different qualitative access to various aspects of modern civilization in areas such as education, health care, global information networks, and the leisure industry. Remaining in a stagnant and marginal social reality produces, and will continue to produce, psychologically and politically conservative thinking, creating conditions that foster xenophobia, isolationism, and retrospective anti-westernism. In other words, under these circumstances the petrol state of Russia cannot be socially balanced regarding the interests of the majority of its people. It cannot be like countries of the developed world.

Why can there not be a compromise between the "beneficiaries" of the petrol state and its "outsiders" on the basis of a fair distribution of the income of exported energy sources? Would it be possible to choose a model that is satisfactory for all groups? In my point of view we cannot look for an ideological answer to this question without trying to explain the inability of Russian society to come to a consensus. This inability is due to the fact that the political elite propagate economic liberalization and minimization of the state's social functions. This mentality of the upper class is a result of the social order established during post-communist transformations. First of all, the modern Russian elite - bureaucrats and businessmen - was formed, under conditions of limited competition, with the help of the government and under its protection. The shift towards new models of social relations requires new equations of social dominance in the higher classes. This also means that the risk exists of losing one's current position. Russian elites are used to solving short-term problems and are not used to taking risks for strategic perspectives. Secondly, due to various factors, the petrol state developed in Russia at a point in history[3] when mass interest in politics and political processes was very low. Society preferred to delegate its right to participate in decision-making entirely to the president, and the demand for opposition activity decreased greatly. In these conditions the autonomy of the higher classes increased. If the political will of the elite is not limited by society and the elite does not see society as a counterbalance to itself, then the ruling stratum will lose any interest in making compromises and determining successions. That way, when two social planes come face to face, no policy is implemented to bring them closer to each other. The petrol state and the elites who control it concentrate all of their forces on keeping the status quo under which they established a monopoly on political power and the main capital of the economy. This circumstance produces the logic of action for the petrol state in fields of policy. This logic is not part of a whole, thought-out strategy, it is implemented on the level of intuition by individuals of the governing elite in actions that are mostly isolated from each other. The social division created by the petrol state and preconditioned by its existence dictates specific reform logic.

No matter how much the apologetics of today's reforms say that they are directed to modernize the state, the reality is completely different. For the petrol state reforms do not attempt to come to higher levels of social and economic organization but are an instrument to optimize the existing system in order to do away with anything that is extra or burdening. We can also add the statement of Mikhail Delyagin: Ridding the government and bureaucracy of social responsibility.[4] The given is that the main criteria of post-industrial

3 During the 1990s the petrol state as a social phenomenon could not exist. Low prices on oil produced the chronicle deficit of money inside the country, in the end made the position of the Soviet elites very unstable and the choice of political courses limited. Secondly the elite's movement towards the desired goal was held back by the strong opposition in the country, the very existence of which evoked fear of returning to the previous social system.

4 См. Делягин М. Социально-экономическая программа будущей революции // Глобальная революция: ретроспективы и перспектива. Материалы к заседанию клуба "Красная палощадь" 13 января 2006 г. М., 2006, p. 3.

modernization is the increasing value of human capital - a goal that is unattainable without expanding access of the latest developments of civilization to the masses: education, health care and science. However, Russian social reforms have a completely different goal, namely lowering budget spending. Other possible aftermaths of reforms - including the fact that significant social groups will lose access to the latest developments of civilization and become marginalized - is not something that worries the authors of reforms. Exactly this happened with the famous law no. 122 - monetization of services, in its latest edition -, when pensioners and other groups receiving aid at the same time lost the possibility to obtain necessary medicines as well as to use public transport in their regions and throughout the country. Only the active protests of such groups forced the government to soften the initial edition of the reform, save some natural forms of aid and allocate extra funds to compensate for the loss of aid. They were thus forced to back away from the goal of their optimization, namely a decrease of budget spending on social aid, which did not occur.

The same optimizing approach is recognizable in the health care reform. For example, one of the reforms goals, the development of health services for the people that is dependent on doctors without specific specializations, shows this approach. For the greater part of the people this will result in the deterioration of the current levels of health care. It is not likely that one of the reforms authors, were he to have health problems, would go to a doctor with no specific specialization! This is a fatal hazard. The beneficiaries of the petrol state are untouched by this reform. The level of their income allows them to consult highly-trained specialists when the need arrives, if necessary outside the country. The ministry of health care's idea of reducing the free hospital stay from two weeks to five days will keep a lot of people from receiving the help that they need.

The same optimization can be seen in the early plans of the education reform. First of all, the petrol state requires professionals with higher education only in a limited number of specializations. From this the desire of administrative bodies emerged to decrease the number of places of higher education, likely merging them with one another (the idea of the minister of science and education Andrei Fursenko to create one big university in Krasnoyarsk on the base of all other institutions was priceless!) in order to lower the number of graduates of other specializations. Secondly, the commercialization of education, including secondary education as well is the fine print hidden in this plan (the proposal, by the same minister, that schools should also be financed by parental sponsorship). At today's levels of corruption, the grants that are awarded to secondary schools and teachers may not reach most of destined schools. Concerning universities and institutions, they are to be made bankrupt like commercial institutions. Thirdly, in the reforms plans the system of education was to be specially altered for the elite, under the criteria that graduates would be drafted to the millitary, with the exception of a privileged list of 35 universities throughout the country where the military department will be retained, but their graduates will not be called up for active military duty. This list contains all of the institutes of higher education where the children of the governing class study (the Moscow State Institute of Foreign Relations, the High-School of Economics, Moscow State University, St. Petersburg University, among others).[5] In this case it is difficult not to agree with the prognosis of the result of these reforms by the former education minister Edward Dneprov: "The new model of education being built will become a means of segregation that will separate education into two sectors: education for 'the elite' and for 'the cattle'." As soon as this happens education will become a social lift that will only lower the larger part of our society. This way a cycle of repeating poorness will be come to be. Children of economically unprivileged parents would never be able to break free of it because their parents would not be able to afford quality education for them.[6]

The real goal of the education reforms, as many specialists believe, are dictated not by care of increasing effectiveness, but to commercialize the vast property of the Russian Academy of Sciences as soon as possible, under the excuse of decreasing budget spending.

Actually, these and similar plans of reforms came into existence during the 1990s. However, because of the above stated reasons, the

5 http://www.utro.ru/articles/2005/09/08/475050.shtml.
6 Эдуард Днепров. "Мы отстаем на две эпохи" // http://www.gazeta.ru/2005/12/16/oa_181971.shtml.

government decided to not to carry them out then. There existed an influential opposition and the position of post-Soviet elites did not seem as determined and their legitimacy in the masses was nominal.

Realization of a long-term reform strategy helps the petrol state resolve an important problem: it restrains social dynamics. These reforms will escalate costs and will lower the living standards in the population. The political significance of this strategy is evident. The society that lived through 10 difficult years of adaptation to a new social order has a constant fear of losing the attained level of stability and will not place high demands on ruling elites. This society will be so preoccupied with adaptation to constantly changing rules of the game that it will not be able to devise long-term social strategies and, hence, question the effectiveness of the ruling powers. It means that the ruling powers need not worry about the emergence of competitors who would claim dominant positions in power structures. In the absence of strong opposition, the society, which is concerned with maintaining certain level of living conditions, will hope that living conditions do not worsen and will rely on existing power structures. Major trends of political development in Russia in the past year confirm this assumption. In population polls, sociologists noticed the public's increased concern with social problems. However, last year, election results for legislative departments in some regions demonstrated that people prefer not to take chances and they give their votes to the existing power and its party, "United Russia."

The middle class is the source of constantly increasing claims and, consequently, of social dynamics in contemporary society. This is why the petrol state is willing to allow a spring tide of prosperity and social ambitions to those groups whose activity is vital for maintaining the normal status of our petrol state. Other groups of population, who cannot aspire to become beneficiaries of the petrol state due to their social or professional positions, encounter an increased cost of living which, in turn, limits the range of their claims and, most importantly, shuts down opportunities for their vertical mobility. The state tries to restrain the interests of those groups. Let us look at some recent examples: In the beginning of 2006, the government adopted new customs regulations which noticeably hardened fiscal pressure on "shuttle trade." An attempt to ban orange and yellow direction lights on automobiles manufactured in America and Japan could substantially hit the pockets of the owners of these automobiles; the government had to abandon this innovation under the pressure of motorists. However, the general vector of government actions in similar situations is becoming very transparent. Since there is no obvious economic motivation behind such actions, the main goal is to restrain the ambitions of certain parts of the middle class. The petrol state is not interested in the well-being and prosperity of the middle class; this is the pivotal difference between the position of the middle class now and its position during the 1990s. At that time, the new elites were not strong enough and lacked confidence; they feared the restoration of communism and required wide political support of a middle class which was just coming into being. This is why the elites were interested in the growth and well-being of the middle class. In the context of the petrol state, there is no threat to the solidity of the elites and there is little need of the middle class as social and political supporter of the elites. Just to the contrary, public sector employees and other state-dependent groups are new partners of the petrol state elites. Sufficient economic resources make it possible to throw "creep-fed rations" to social groups at the lowest levels of the social ladder. An important statement of the Finance Minister Alexej Kudrin was very symptomatic in this respect. Kudrin announced that in 2008 the country would "live better than it lived in that historical period which doctors, teachers and people of older generation remember." Thus, the political measuring stick (precisely political, because we are speaking about the social memory of people) is the 1990 rate of consumption of people paid from the state budget. Attainment of this rate is supposed to warrant stability of the petrol state.

Thus, the ideal social structure for elites of our petrol state is the one where the irremovable minority deeply integrated into the world – the elite – would reign supreme. The most numerous groups of population at the lowest levels of social ladder, who would be loyal to this reign, would constantly adapt to changing conditions. Between these two, there would be an insignificant layer of middle class with no influence whatsoever. Similar social structures are characteristic of many countries of Latin America of the second and third order of development (e.g. Bolivia and Peru). Their structures are characterized by stability, and societies based on them are characterized by a steady inability of internal evolution.

The conclusion that follows is simple. Reforms in the realities of the petrol state will inevitably promote the existing system and will restrain social dynamics since the petrol state is predominantly oriented to self-preservation and is trying to minimize risks wherever possible. I should note here that this essay did not consider political aspects related to democracy problems in Russia. That was done intentionally. Let us assume that those who believe that realities of the petrol state are sufficient for modernization of the country are right, and that there is no need for democratic institutions. However, even if we abstract from political constituent of the problem in question, we will have to admit that the social facets of these realities lack intrinsic sources of development. Existing elites feel very comfortable in the grand scheme of things and would not bother with modernization projects, especially when risks are involved. Economic and administrative restrictions constrain the activities and fervour of other social groups. Should Russia's ruling powers decide to devise and implement a modernization project for Russia, they must clearly understand that the implementation of such project is only possible if the petrol state and the public system created by it are dismantled.

Vladimir Milov

The Role of Oil & Gas in the Russian Economy

This paper briefly examines the key factors of influence of the Russian oil & gas sectors on the country's economy as well as the current status of the oil & gas industries in Russia in order to give a general idea on how dependent the country's economy is on oil & gas, and how sustainable this dependence is, given the known prospects of development of the oil & gas industries in the future.

A General Overview of Russia's Dependence on Oil & Gas

The role of oil & gas in the Russian economy is, on one hand, often overestimated in public. On the other hand, this role remains vital in certain aspects. This importance is primarily related to a high dependence of economic growth on the growth in oil & gas export volumes and/or prices, as well as the high extent of dependence of the federal budget on oil & gas revenues. In general, it should be said that oil & gas play only a limited role in the structure of the Russian economy (as far as the share in GDP is concerned) – but, at the same time, they are one of the few competitive sectors driving the country's economic development, and they remain the strongest foundations behind Russia's fiscal balance and macroeconomic stability as well.

In the overall structure of Russian GDP, the role of oil & gas industries is not large – according to official estimates, oil & gas industries contribute about 9 per cent to Russian GDP. However, it should be noted that such a small share of oil & gas sector in Russia is debated by some economists, who argue that in reality, if the transfer pricing in the oil & gas industry is taken into account, the real share of oil & gas industries in GDP can be estimated as being 25–30% (Ruehl, Schafer, 2004[1]). The transfer pricing is a scheme in accordance with which the oil producing subsidiaries of the large oil & gas holding companies resell the produced oil to the trading subsidiaries of the same holdings, under prices regulated by the holding corporation itself, which are set normally below fair market value of the commodity with purpose to redistribute part of the value to trading subsidiaries. Under this scheme, the income of the oil producing companies is underestimated, whereas part of the added value produced by the mineral extraction sectors is wrongly attributed to the "trade" segment of the economy, proponents of this theory argue.

This theory seems to have strong foundations in reality; however, due to the poor quality of Russian statistics and the methodology of the data collection, it appears virtually impossible to identify how much added value is actually being wrongly attributed to the trade segment of the Russian economy. The sources mentioned above largely rely on estimates of the real share of oil & gas in the GDP, whereas the actual extent of influence of the "Potemkin GDP" factor remains unknown.

But, nevertheless, even if the transit pricing factor is concerned, and the share of petroleum industries in the Russian GDP is considered to be above 20%, oil & gas still do not play the dominant role in the Russian GDP, thus not allowing the characterization of Russia as a monoproduct economy similar to other petro-states. It should be said that other manufacturing sectors as well as services account for a significant portion of Russian GDP; in this way Russia is very different from other, more traditional petro-states, in which the petroleum industry normally comprise 70% of GDP or higher.

However, oil & gas play very important roles in two major issues crucial for Russia's economic sustainability and development: economic growth and budget revenues. As for economic growth, it is quite obvious that it was the oil & gas exports that were the major factor behind Russia's impressive economic growth since 2000, with average annual GDP growth rates being in the range of 6–7%.

1 Christof Ruehl, Mark Schafer, "Potemkin GDP", Vedomosti, February 19, 2004; more details in the World Bank Russian Economic Report 7, February 2004.

Before 2005, growth in the physical volume of exports and the direct effect of growing international oil prices had contributed most significantly to the GDP growth; in 2005–2007, despite the fact that the growth of oil & gas production had substantially slowed down, and international oil prices had more or less stabilized within the range of $ 55–70 per barrel, growth continued due to an increase in domestic consumption fuelled by the high levels of oil & gas export revenue inflow. The estimated decomposition of factors contributing to economic growth in Russia, as suggested by the Moscow-based Center for Macro-Economic Analysis and Short-Term Forecasting, is shown in Table 1.

Table 1: The Factors Contributing to Economic Growth in Russia in 2000–2007

	2000	'01	'02	'03	'04	'05	'06	'07 (est.)
GDP growth rate in %	10.0	5.1	4.7	7.3	7.2	6.4	6.7	7.5
of this attributed to:								
Growth in physical volumes of commodity exports (primarily oil & gas)	2.8	1.6	3.0	3.0	3.4	1.5	1.9	1.9
Growth in international oil prices	4.9	1.0	-0.7	1.6	1.7	2.0	1.0	1.4
Other factors (non-tradable services, domestic consumption, etc.)	2.3	2.5	2.4	2.8	2.1	2.9	3.8	4.3

Source: Center for Macro-Economic Analysis and Short-Term Forecasting (CMAKP), www.forecast.ru. Numbers may not add up due to rounding.

As the table shows, in 2000–2005, economic growth driven by "other" factors (other than growth in oil & gas export prices and physical volumes of exports) comprised only 2.5% of annual GDP growth in average. This is where the Russian economy probably would have been if the growing oil & gas exports had not grown considerably and international oil prices (and European gas export prices, which are fundamentally linked to international oil prices with a lag time of 6–9 months) had not increased. The dynamics of these factors are shown in Charts 1 and 2.

As can be seen from the charts, the physical volumes of Russian oil & gas exports as well as international oil & gas prices largely grew steadily until 2004 (with the exception of 2002, when the oil & gas prices fell substantially, contributing to the reduction of the GDP growth rate to 4.7%, the lowest level in the 2000–2007 period).

In 2005–2006, the rates of growth of oil production and exports slowed down substantially: the oil industry was apparently suffering from the escalation of interventionist trends in the state economic policy towards the oil sector (see below). Starting in 2005, a key contribution to economic growth was made by the growth in domestic consumption, as can be seen from Table 1 above.

However, the large oil & gas revenue windfall continues to be the major factor that supports growing domestic consumption, and no guarantees of the sustainability of growth in consumption remain if it does not continue to be fed by surging oil & gas export revenues.

It thus appears that the impressive Russian economic growth of 2000–2007 can be largely attributed to the performance of the oil & gas industries; without them, economic growth would most likely have been much more modest, with GDP growth rates hardly exceeding 2–3% per year.

The role of the oil & gas industries is also crucial to federal budget revenues. According to the federal budget for the years 2008–2010[2] developed by the Russian Ministry of Finance, oil & gas revenues will continue to play an important role in the Russian federal budget for the coming years, although the Russian Ministry of Finance (Minfin) expects that this role will also continue to decline. The Russian Minfin projects that oil & gas revenues of the federal budget will decline to 5.4% of GDP in 2010 as opposed to the 8.2% that were expected in 2007, and to just 29.6% of the total revenues of the federal budget, as opposed to the 40.2% of 2007 (see Charts 3 and 4).

However, it should be said that Minfin's projections were based on a very conservative forecast of international prices and export

2 Approved by the Russian federal law of July 24, 2007, 198-FZ.

Chart 1: Russian Oil & Gas Exports in 1999–2006

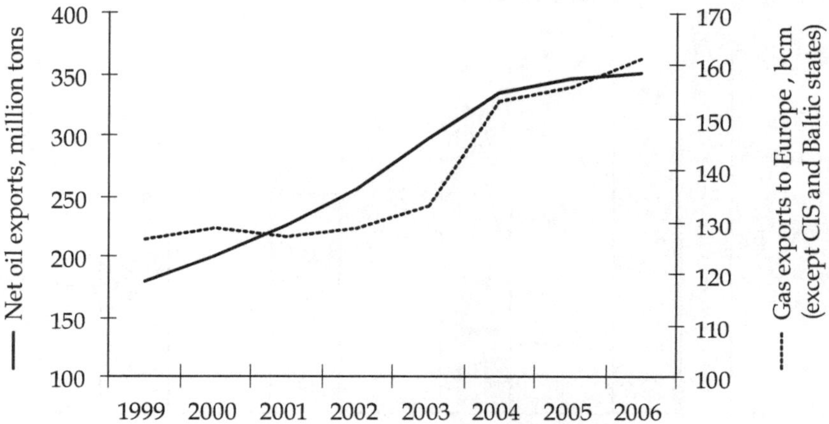

Chart 2: Russian Oil & Gas Export Prices in 1999–2006

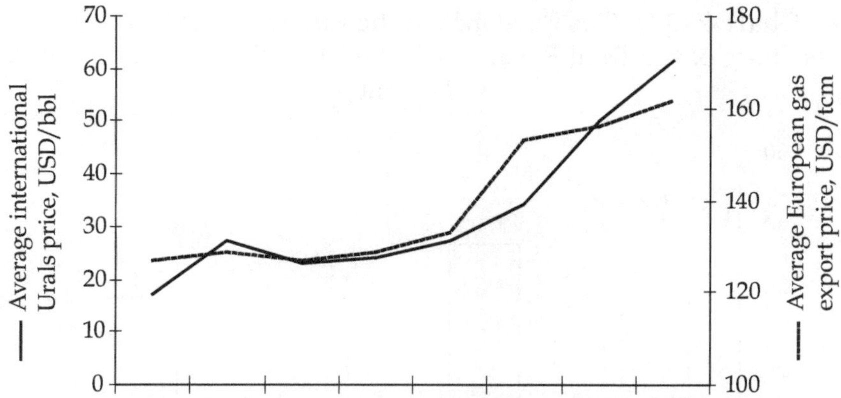

Source: oil prices – The Wall Street Journal, gas prices – Gazprom (only available starting from 2001).

Chart 3: Oil & Gas Revenues of the Russian Federal Budget as Share of GDP, 2007–2010, in Percent

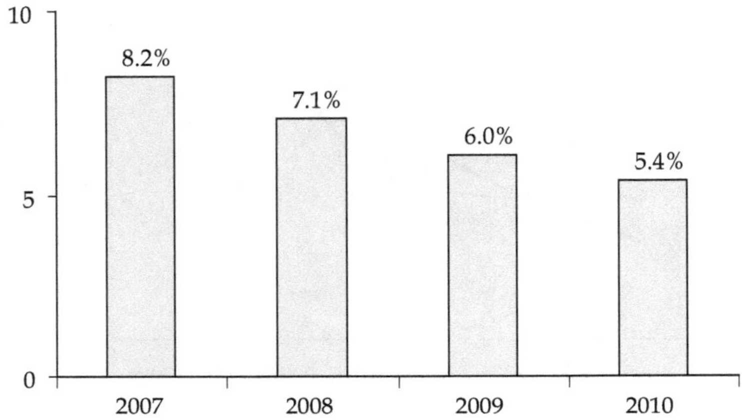

Source: Russian Ministry of Finance, supplementary documents to the federal budget for 2008–2010.

Chart 4: Oil & Gas Revenues of the Russian Federal Budget as Share of the Total Revenues of the Federal Budget, 2007–2010, in Percent

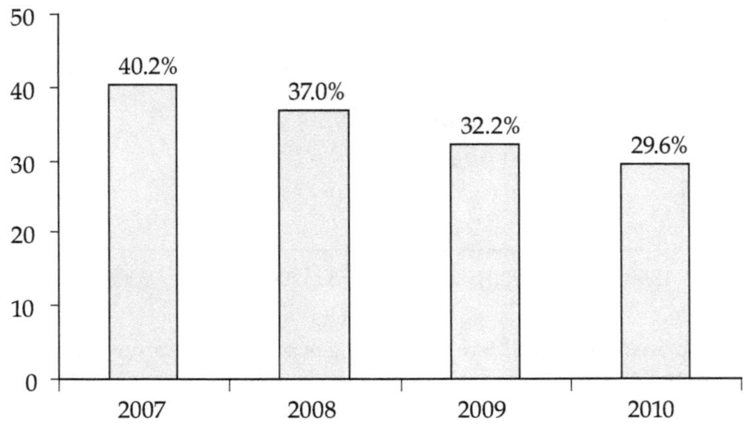

Source: Russian Ministry of Finance, supplementary documents to the federal budget for 2008–2010.

growth rates of oil & gas. Minfin projected that the average Urals price would amount to $ 53/bbl in 2008, $ 52/bbl in 2009, and $ 50/bbl in 2010; the average price of natural gas exported to Europe (except CIS and Baltic states) would be $ 247.6/tcm, 224.7/tcm, and 219.6/tcm in 2008, 2009, and 2010, respectively. The average growth of exports of oil and petroleum products was expected to be 1.3% per year in 2008–2010, as compared to an average of 3.6% in 2005–2007. (It should be considered that, for instance, oil production in Russia grew by 2.6% during the first 8 months of 2007; most of the additional oil was exported.)

The projected growth in Russian gas exports to Europe (4%) was relatively high, but gas exports in Russia are much less significantly taxed than oil exports (for instance, the only gas exporter Gazprom paid only $ 15 billion in taxes in 2006, or just 6.5% of the total federal budget revenues - most of the revenues come from the oil sector and not the gas sector), which means that the growth in oil exports remains a much more significant factor of influence on budget revenues than the gas exports.

Also, it appears that Minfin was probably too optimistic in its forecast that the growth of the domestic manufacturing sectors in Russia would displace imports. As Minfin projected, the growth in overall imports would decline from an average 19.8% a year in 2005–2008 to 15% in 2009 and just 12.4% in 2010.

Such an assumption may in reality mean that the share of the oil & gas industries' contribution to GDP growth and federal budget revenues would in fact decline and be replaced by domestic products. However, there are no reasons to believe that this would be a particularly likely scenario, taking into account the fact that domestic manufacturers in Russia are strongly suffering from the "Dutch disease" and are often at a competitive disadvantage against importers due to the appreciation of the rouble. In this regard, the projected switch from imports to domestic manufacturers (and the growth of share of the latter in Russian GDP and budget revenues) does not appear particularly realistic.

All this allows us to expect that, in fact, the oil & gas revenues of the federal budget will most likely stay above 40% of the total revenues of the Russian budget in the coming years, or not less than 7.5–8% of the Russian GDP.

It is also important to stress that Russian Minfin in its recent calculations related to the federal budget of 2008–2010 admitted that, without oil & gas revenues, the Russian federal budget would have deficits of 6.6% of GDP in 2008, 5.9% of GDP in 2009, and 5.3% of GDP in 2010. (As said above, this deficit will be even larger if the share of oil & gas revenues in the budget turn out to be higher than expected.) This actually means that, had the international oil & gas prices halved from the present levels, the Russian budget would have had a deficit of around 3–3.5% of GDP in 2008.

It should be also said that Russia has accumulated a Stabilization Fund from the oil & gas revenues, which is kept on the Government's accounts in the Russian Central Bank with the purpose of covering the potential future risks of oil & gas export revenue decline (although, it should be said, the goals of accumulating the Stabilization Fund are not absolutely clearly defined – it was launched mainly as a "defensive" project, aimed at sterilizing excessive oil & gas revenue windfall in order to combat the "Dutch disease" and inflation risks, not as a fund aimed at some clearly understood future goals). As of September 1, 2007, according to Minfin, the amount of money accumulated in the Stabilization Fund had reached $ 132.9 billion, or about 11.1% of the Russian GDP.

The Stabilization Fund can theoretically act as a "safety pillow" for the Russian budget should the international oil & gas prices fall substantially in the future. However, it could to support the current economic model for a period no longer than 3–4 years, after which the future of the Russian economy would remain uncertain. Also, there is no guarantee that growing federal spending would not eat up large part of the Stabilization Fund before such a crisis occurs: the non-interest spending of the Russian federal budget has skyrocketed from 12.3% in 2004 to a projection of 18.3% for 2008–2009, and further growth can be expected.

Challenges Facing Russia's Oil & Gas Industries

When discussing the oil & gas sector's role in the Russian economy it is important to understand the internal developments in this sector since these may determine the future abilities of this sector to assist in the further economic development in Russia.

Russia is one of the world's leading energy producers and exporters. There is good reason to believe that even the official figures of the country's oil & gas reserves seriously underestimate Russia's resource potential; large perspective provinces (Eastern Siberia, North Ocean and Pacific offshore) are underexplored. But, even taking into account the official numbers, it is quite evident that Russia will remain a large producer and a net exporter of major energy resources for decades.

Table 2: Russia's Energy Potential as of 2006

	Oil	Gas
Proved reserves, mtoe	10,900*	42,885
Share of global proved reserves,%	6.6	26.3
Indigenous production, mtoe/year	480.5	550.9
Share in global production,%	12.3	21.3
Net exports, mtoe/year	351.5	161.5
Reserve to production ratio	22.3	77.8

* Russian official estimates show 22,000 mtoe; there are fundamental differences in reserve classification methodology and difficulties with access to data.

Source: BP Statistical Review of World Energy 2007.

However, despite her vast energy potential, Russia continues to experience major challenges in terms of the strategic future of the country's energy sector, due to uncertainties with regard to the state's energy policy model and an unfavourable structural situation within the Russian energy sector. These present serious obstacles for successful and efficient development of the energy sector as well as the perspectives of international integration of the country's energy sector.

During the past 15 years, the Russian energy sector has clearly evolved in two unequal parts: one part of the energy sector has undergone major market restructuring and has been liberalized and privatized to a large extent (the oil sector), while another part still remains within a Soviet-type centralized structure, under price regulation and state ownership (the gas sector, which largely remained under the centralized control of Gazprom). The economic situation and the prospects for further development in these two segments of the energy sector sharply differ. While the privatized and liberalized oil sector has clearly succeeded during the last decade in strong recovery growth efforts and has dramatically increased in economic efficiency, the centralized and state-run gas sector continues to experience tremendous problems with underinvestment in the core businesses, inefficiency and non-transparency (which add up to the inability to mobilize necessary investments), and growth in production.

For instance, oil production increased by 57% from 1999–2006, a result far exceeding the expectations of the best oil sector experts (oil production by the state-owned oil companies, on the contrary, grew only by 14%) – this being the sector which in the early 1990s had received subsidies from state and international financial institutions for the purpose of rehabilitating oil production! This result was achieved mainly due to massive capital investments by private oil companies (between 2001 and 2004, an average of more $ 6 billion annually were invested in upstream oil production, according to Rosstat), and a significant increase of productivity of oil wells (by an average of 37% in 2005 as compared to the level of 2000, from 7.5 to 10.3 tons per day per average oil well). This is the result of the introduction of modern Western technologies to the Russian production upstream and is due to the hiring of the best international oilfield service companies, such as Schlumberger and Halliburton. Private Russian oil companies also opened new oil fields, including the large new Priobskoye oil field in Western Siberia introduced by the privately owned Yukos oil company, which produced insignificant volumes of oil in the late 1990s but reached a peak production of over 20 million tons of oil per year in 2004 when it was been taken over by the state-owned oil company, Rosneft.

As a result of this successful development, the share of Russian oil production in the global oil production increased from 8.8% in 1999 to 12.3% in 2006.

Private initiative has clearly been successful in the gas industry as well: in the small segment of gas production occupied by independent private gas producing companies, gas production in 2006 more than doubled when compared to levesl of 1999 (106 bcm per year as compared to just 47 bcm in 1999), despite blocked access to exports and primarily greenfield projects requiring substantial investments.

In the state-controlled Gazprom, the picture was quite different. Gas giant Gazprom, until recently, was hardly able to even reach 1999 gas production levels – the cumulative gas production in 2006 had grown by just 2% as compared to 1999, even taking into account the state-backed hostile takeovers of several independent gas companies like Purgaz and Northgas (without these hostile takeovers, Gazprom's gas production in 2006 would have been 2% less than that of 1999).

Chart 5: Cumulative Growth in Oil & Gas Production in Russia in 2006 as Compared to 1999, in Percent

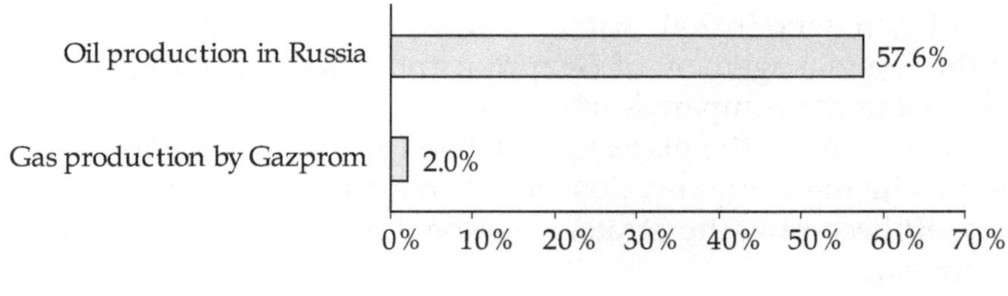

Gazprom, unlike the oil companies, has been showing disappointing results both in upstream production development and in efficiency. This is even more disappointing, as the gas industry had approached

the beginning of the 21st century in much better shape than the oil sector: the output decline in the 1990s was not severe and the gas producing fields were relatively new when compared to major oil fields. However, unlike the oil companies, Gazprom failed to invest enough in the development of the upstream gas production sector; this is the root cause of the stagnation in the gas production sector and led to a very clear perspective for a severe gas production decline in the future: average annual capital expenditures in upstream gas production accounted for just $ 3.4 billion in 2001–2007, only half as much as in the oil sector. The average productivity of gas wells in Gazprom declined by more than 13% between 2001 and 2006, from 98 million m^3 per well per day to just 85 million m^3.

Initially, the first Russian government under President Vladimir Putin (headed by Mikhail Kasyanov) in 2000 announced ambitious plans for the market restructuring and privatization of Gazprom in order to make Russia's gas industry more transparent, efficient and competitive in global terms, under the same model of development as that of the liberalized and privatized oil sector.

There were also a number of promises to encourage more direct foreign investments into the Russian oil & gas sector, and privatize the remaining oil producing companies owned by the state.

Unfortunately, these reforms had to a large extent slowed down by 2002–2003, being replaced by policy trends of a reverse nature that had started to emerge since 2003–2004. The implementation of a market restructuring plan by the centralized gas monopoly Gazprom (which was developed in 2002) was banned by President Putin, who in 2001 had appointed very close affiliates of his to the top management of Gazprom and has been acting as a key lobbyist of the company's interests ever since. In 2002, the government suspended the plans to privatize the last major state-owned oil producing company, Rosneft; at that time the rumours about Rosneft becoming the 'strategic national oil company' had started to appear.

In early 2003, Russian authorities openly stated that they would not tolerate the appearance of a privately owned oil pipeline infrastructure in the country, preventing the implementation of the Western Siberia-Murmansk oil pipeline project initiated by four private Russian oil companies (Yukos, Lukoil, TNK-BP and Sibneft) and targeted to assist the development of oil production in the regions of Western Siberia and Timano Pechora by creating a new channel of export of oil to the international markets, including the United States (the Murmansk area in North-Western Russia is the only coastal area located within reasonable distance from main oil production regions that would allow the creation of a deepwater oil export terminal, at the same time being a non-freezing area, unlike the Baltic or the rest of the Russian Arctic coast). The project was banned because state authorities were unwilling to accept the ownership claims of the private oil producing companies. If built, the pipeline would obviously have become the best Russian oil export channel, thus challenging the current oil pipeline export system owned by the state. The abandonment of the Western Siberia-Murmansk oil pipeline project delayed development of new oil production in Western Siberia and Timano Pechora, which would have been boosted if private oil producing companies would have been allowed to create a new channel of exports of additionally produced oil. Instead, the Russian Government made a decision to build the subsidized oil pipeline from Eastern Siberia to the Pacific, the construction of which had just began; this will hardly provide private oil producers with competitive export alternatives.

In 2003–2005, more negative developments occurred, demonstrating the will of Russian authorities to pursue more state-interventionist policies in the energy sector and to throw aside the liberalization agenda declared in 2000. These developments included:

1) The Yukos affair,[3] which by mid-2004 had clearly become an asset redistribution deal rather than the political case of Yukos ex-CEO Khodorkovsky, as it seemed in the beginning, and had turned into the destruction and forced takeover of the company's assets, significantly slowing down oil production by major Yukos subsidiaries.
2) Taxes for oil production and exports were severely increased in 2004 after the parliamentary and presidential election campaign.

3 This did not end with the hostile takeover of Yuganskneftegaz in December 2004; further tax and other criminal charges were brought against Samaraneftegaz and Tomskneft, the largest remaining Yukos upstream subsidiaries.

Taxes on private oil companies were increased to $ 31–34 per barrel, leaving oil producing companies with just $ 26–29 of revenues and virtually eliminated the economic motivation for companies to develop marginal oil fields even during a time of exceptionally high international oil prices.
3) The state's control equity stake in Gazprom was restored.
4) The acquisition of some private oil companies (Yukos, Sibneft) by state-dominated majors like Rosneft and Gazprom, resulting in a rise of the state's share from 15% in 2003 to 40% in 2007; more acquisitions of private oil companies are being discussed.
5) The hostile takeovers of the independent gas producing companies by Gazprom (Northgas, Sibneftegas).
6) The intrusion of Gazprom into the 100% privately and foreign owned Sakhalin-2 project.
7) Increased Kremlin interference in approval of large direct foreign investment deals related to Russian energy companies.
8) The law on monopolization of gas exports from Russia into the hands of Gazprom adopted in July 2007.
9) Restrictions to foreign participation in oil and gas upstream were announced.

These developments may not necessarily represent a total restoration of direct administrative control over the energy sector of the economy (private initiative is still tolerated to a certain extent). It would also be not exactly correct to expect that Russian authorities are necessarily moving towards establishing a single national oil and gas enterprise, one company incorporating all the oil & gas assets controlled by the state. There are credible reports of a major rivalry among the key influence groups inside Kremlin for control over various oil & gas assets: it would thus be reasonable to expect that we are entering an era of permanent asset redistribution under bureaucratic control rather than an era of structural stability with some strategic goals.

But it is quite clear that Russian authorities have completely reversed the liberalization and deregulation agenda which they had announced in 2000, not only regarding the energy sector, but as a general principle of economic policy (some interventionist developments of the same sort are happening in other sectors as well – banking, machine-building, etc.).

In this environment, an important issue is the access of foreign investors to Russian energy upstream sectors and infrastructure. In an address to Russian Federal Assembly (Parliament) in May 2005, President Putin openly said that in some "strategic areas" of the national economy, opportunities for foreign investments should be restricted. This most definitely applies to the energy sector, as the intention to restrict foreign equity participation in oil and gas field operators was announced earlier, during the development of the new subsoil legislation in 2004 and early 2005.

All in all, Russian policy towards foreign investments in the energy sector seems somewhat eclectic and confusing. On one hand, a number of companies have received clear approval from the Kremlin for direct equity investments in Russian oil and gas business (BP, ConocoPhillips, BASF, E.oN-Ruhrgas and several others are negotiating projects with Russian companies). To some it may seem that this is just a matter of taste of the Russian authorities, and companies just need to pass the "loyalty test" to be allowed into Russian upstream. For instance, experts often refer to the example of ExxonMobil and Total companies, which are widely believed to have failed in their attempts to make direct foreign investments in Russia because of their inability to find the "right individual approach" to the Russian leaders.

The truth is, however, much more complicated. First, it should be said that the Kremlin does its best to avoid setting up clear rules and conditions for foreign access to subsoil operations. Even the new subsoil legislation reserves the right of authorities to decide on access permissions on a case-by-case basis, without any transparent criteria or conditions. This creates quite a vulnerable environment even for the "approved" investors, as the conditions may suddenly change according to the state of mind of the authorities. Second, even the "approved" deals show that the Kremlin is only ready to tolerate minor foreign equity participation in oil & gas field operating companies, at least below 50%, but in most cases well below 25%. Taking into account the very non-transparent corporate governance practices of most Russian state-dominated companies, very little room is left for actual control over operational and investment decisions by foreign investors in these joint ventures; for them, more than exclusive global marketing opportunities do not remain. It is questionable whether this "cooperation concept" will be beneficial for foreign companies in a long run and to what extent it will succeed.

As a complex result of these new policy developments, oil production growth rates in 2007 were down to 2.5% a year from an average of 8.5% a year in 2004, when, in fact, a number of oil industry experts had predicted oil production growth in Russia of 4–5% a year would continue for at least 4–5 years before a slowdown. It should be said that in 2006–2007 the international oil companies were the drivers of growth – operators of the production-sharing agreements (PSA), greenfield projects with largely foreign participation that had been opened just recently, started to deliver major production volumes in 2005–2007. These projects include the Sakhalin-1 project led by ExxonMobil, the Sakhalin-2 and Salym projects led by Shell, and the Kharyaga project led by Total.

In the rest of the oil production sector, growth decreased: if the growth of oil production by PSA operators is not accounted for, Russian oil production would have grown only 1.6% in 2006, and 0.4% during the first 8 months of 2007, as compared to the actual 2.2% and 3.0%, respectively.

However, foreign investors, who have contributed most even to the modest growth of oil production of the recent two years, were seriously attacked in the recent period: authorities have restricted (mostly through informal mechanisms) their access to development of the "green" oil fields, and in the currently operated projects their positions have been severely challenged. Gazprom, through the use of administrative pressure on the Sakhalin-2 project, managed to force the foreign investors (Shell, Mitsubishi and Mitsui) to sell their control equity share in the project. After that, TNK-BP (a joint venture between BP and the Russian oil company TNK) was forced to sell to Gazprom an equity share of the Kovykta gas field in Eastern Siberia. The Sakhalin-1 project led by ExxonMobil has been threatened with a potential blockade of its gas export contract with China, as it competes with Gazprom's efforts to sell gas to the Chinese market. A large number of experts believe that either the takeover of the Sakhalin-1 project, or a forced transfer of gas exports marketing rights to Gazprom, are likely to take place in the future.

In this environment, when private Russian oil & gas companies and international investors are largely being banned from control over major new projects, the only players who can potentially develop the complicated, remote and underdeveloped new "greenfield" production areas without developed infrastructure – the Arctic and Pacific offshore, Eastern Siberia, the Yamal Peninsula – remain the state-owned energy giants Gazprom and Rosneft. However, due to their inefficiency, and after the acquisitions of privately owned assets, these companies have accumulated huge debts which impose constraints on their ability to timely develop new oil & gas provinces. For instance, the debt to revenue ratio at Gazprom and Rosneft reached 0.5 and 0.4, respectively, at the end of 2006, and has probably reached 0.6 and 1.0 by now, after recent acquisitions of 2007 (the most recent financial reports are not yet available), compared to the 0.02–0.15 debt to revenue ratio that most large international oil & gas companies have. Gazprom's debt have reached approximately $ 50 billion by now, and Rosneft's about $ 35 billion. This obviously impairs the ability of these companies to invest in new oil & gas production fields.

At the same time, the potential of oil & gas production in mature fields is gradually expiring as they reach depleted stages, and the country's oil & gas production profile mostly depends on development of the new "greenfield" provinces. But the companies who are potentially able to do that (Russian private and international oil & gas companies) have been either destroyed or restricted from participation in new projects, and the state-backed energy giants Gazprom and Rosneft are unable to develop these fields both due to their huge accumulated debts and the lack of experience, advanced technologies and skills (particularly with regard to development of offshore oil & gas fields).

The situation with the decline of output at mature fields is the most severe in the Russian gas industry. As a matter of fact, Table 3 shows that two major matured Gazprom's gas production fields, Urengoyskoye and Yamburgskoye, have lost around 140 bcm of annual gas production between 2000 and 2006 and are expected to lose more production in the coming years.

Table 3: Gas Production at Gazprom's Matured Fields Urengoy and Yamburg, 1996–2020 (bcm)

	1996	2000	2006	2010*	2015*	2020*
Urengoyskoye	242.2	193.3	109.5	96	70	52
Yamburgskoye	176.5	168.0	120.0	104	77	54

* projection

Source: Jonathan Stern, "The Future of Russian Gas and Gazprom", Oxford University Press, 2005; Institute of Energy Policy data.

Considering the necessary lag time of 5 to 7 years between the initiation of a major development and the beginning of production at large remote fields such as the Yamal Peninsula fields (Bovanenkovskoye, etc.), no other super-giant gas fields can be realistically brought onstream before 2015–2017, which suggests that Gazprom currently faces a strong potential gas supply deficit challenge.

Recent attempts to establish equity partnerships with international oil & gas companies in order to develop Russian upstream gas fields have not proven to be particularly successful. Several years ago, after Russia had started to formally and informally limit foreign access to development of its oil & gas resources, Russian authorities suggested that certain international energy companies could be granted access to upstream Western Siberian gas production assets as a part of the widely praised "asset swap" scheme, in accordance with which international energy companies were allowed to swap equity shares in their worldwide energy assets in return for access to upstream energy production assets in Russia.

Subsequent developments with the implementation of the "asset swap" scheme in relations between Gazprom, however, revealed the limitations of this scheme, demonstrating the particular non-transparency of this process and inequalities in Gazprom's treatment of its international partners. Whereas Gazprom has been seeking full-graded access to downstream energy assets in Europe as a part of the "swap" scheme, it had been agreeing to provide only symbolic, rather insignificant "access" to its own downstream assets for its international counterparts. Only one international company, the German company Wintershall, received equity shares in two relatively minor projects – a 50% share in the Achimgaz joint venture in Western Siberia, a greenfield gas project with a peak gas production that should reach just over 8 bcm per year, and 25% minus one share in the operator company of the South Russkoye gas field in Western Siberia with an expected peak gas production of 25 bcm per year (which is yet to be acquired). The negotiations to grant another German company, E.oN-Ruhrgas, access to the Western Siberian upstream gas production assets are not yet successful and are proving to be extremely difficult. Negotiations on asset swap between Gazprom and Shell (Gazprom was supposed to swap the participation in the Western Siberian upstream assets in return for acquiring control over a stake in the "Sakhalin-2" project) have failed.

All the gas that is to be produced at the fields where foreign companies had already acquired or are yet supposed to acquire an equity share, must be sold to Gazprom immediately on the wellhead basis at the prices unilaterally set by Gazprom. Neither direct marketing nor exports of gas by the joint ventures at market prices are allowed; this is the standard Russian practice in the environment of the total and unchallenged Gazprom's monopoly over the trunk gas pipeline infrastructure. In fact, this practice was strengthened last year due to the adoption of the "gas exports" law in Russia in July 2007, which legally banned the possibility of gas exports by any company except Gazprom.

In such an environment, foreign counterparts of Gazprom have virtually no rights for the marketing of gas produced; they cannot sell this gas on international markets and are forced to sell it to Gazprom under prices that cannot be negotiated under fair conditions. The role of foreign companies in the upstream development of the Western Siberian gas fields thus remains quite limited. Such a situation demonstrates the unequal nature of the positions of Gazprom and its German counterparts in their energy relations, and the selective and non-transparent nature of the current regime of "access" to the Russian upstream energy assets.

Recently, the French oil & gas company Total has apparently been granted "rights" to participate in the development of the massive Shtokman gas field in the Barents sea; in reality, however, this is a

25% minus one share participation in a "special purpose company" which will have neither legal rights for field development nor the rights to the produced gas and its marketing (the status of the "special purpose company" must be, in fact, quite similar to that of a service company) and does not provide Total with the rights to take part in decision making on field development schedule and technical solutions, gas sales, and revenue sharing. In this respect, it is yet unclear whether an announced "partnership" between Gazprom and Total in development of the Shtokman gas field will prove successful.

It is quite questionable whether these examples will both prove the efficiency of the scheme of "minor" participation of foreign investors in oil & gas projects in Russia led by state-owned companies, as well as serve as encouraging examples for other potential international investors in the Russian oil & gas "greenfield" projects in the future. Low chances for large-scale participation of international investors in the Russian oil & gas sector upstream allow the prediction that the development of the new "greenfield" oil & gas production areas may be, at large, seriously postponed, which will increase the risk of a decline in Russian oil & gas production in the mid-term future.

Such a development would present a serious challenge to the sustainability of Russia's future development, a large part of which currently relies on the contribution of the oil & gas industries.

Nina Poussenkova

In Search of "Oildorado" …

East Siberia and the Far East: Russia's Petroleum Frontier

Until recently, the double-headed eagle of Russia was gazing in the western direction. Historically, Europe, linked to its eastern neighbour by oil and gas pipelines, has been Russia's main trading partner in the energy sphere.

During the 1990s and early 2000s, Russian oil and gas companies made sporadic efforts to diversify their activities beyond Europe, but these attempts were not fully supported by the state, whose energy strategy was mainly focused upon Europe.

However, the problem of strengthening Russia's energy security has currently become one of the top priorities of the RF government, which is now developing the eastern vector. The "turning east" initiative is also determined by the urgent need to strengthen the national security of Russia through revitalization of East Siberia and the Far East, including through the development of their huge untapped hydrocarbon potential.

In terms of strengthening her energy security, Russia intends to diversify the export destinations of Russian oil and gas, which are now predominantly focused on the stagnating European markets, by establishing a foothold in the rapidly growing Asia-Pacific countries and accessing US consumers. Geopolitically, Russia also needs to develop a new relationship with China, since there is now serious concern that Russia's East may turn into a "resource hinterland" for its southern neighbour.

The key question is whether Russia will be able to repeat the unprecedented accomplishment of the 1960s and 1970s, when the USSR launched the West Siberian petroleum province. The difficulties connected with the development of East Siberia and the Far East will be even more challenging financially, economically, technologically, and socially.

Meanwhile, Russia no longer has the command-and-control mechanisms that it had at its disposal under Socialism and that it fully took advantage of in order to create the West Siberian petroleum province. Though the urgent need to revitalize the East of Russia was already apparent in the 1990s, during the turbulent transition period Russia did not have enough money to undertake such a daunting task, and Moscow did not exercise sufficient control over the regions to implement a coordinated effort required to bring the new petroleum province on line.

In the current decade, the situation has radically changed. On the one hand, the high world oil prices and the petroleum revenues the country is generating make the project more achievable financially. Besides, greater economic and political stability in the country and much stronger authority of the Kremlin vis-à-vis the Federation subjects seem to be important prerequisites for creating a new petroleum province and entering the Asian oil and gas markets.

However, the delays in the implementation of key projects in the East of Russia, such as the construction of the East Siberia–Pacific Ocean pipeline and commissioning of the LNG plant of the Sakhalin-2 project, raise doubts about the ability of Russia to achieve real progress in the development of the eastern petroleum frontier in the foreseeable future. Russia's leadership, which is strongly focused on using the windfall profits generated by high world oil prices to meet its short-term goals, is either unwilling or unable to apply market mechanisms by providing appropriate incentives to domestic companies that are ready to work in the East of Russia and to involve foreign majors in civilizing these regions.

1. The Hydrocarbon Potential of East Siberia and the Far East

Currently, there are four oil and gas provinces in Russia (Table 1):

Table 1: Distribution of Total Initial Oil and Gas Reserves by Region (in Percent)

Region	Oil reserves	Gas reserves
West Siberia	53.5	41
East Siberia and the Far East	13.5	19
Continental shelf	12.4	32
European part of Russia	20.6	8

Source: Oil and Gas Vertical (OGV), 17, 2005, p. 35.

East Siberia and the Far East have less oil and gas reserves than West Siberia, but nevertheless belong the few remaining untapped petroleum provinces in the world.

The reserve estimates for East Siberia and the Far East range quite widely. The Siberian Division of the Academy of Sciences projects that recoverable resources amount to 20–22 billion tons of oil and 58–61 trillion m^3 (tcm) of gas. Academician Alexei Kontorovich cites 15–20 billion tons of oil, 35–40 tcm of gas and 3–4 billion tons of condensate.[1] The lion's share of hydrocarbon resources is located in remote Arctic regions and will not be in demand for the next 20–30 years.

Production level estimates also vary enormously. The *Energy Strategy of Russia up to 2020* forecasts that oil production in East Siberia and the Far East could reach 3 million tons under a critical

1 Recommendations of the parliamentary hearings "Legislative Basis for Creating Pipeline Systems in East Siberia and the Far East", Moscow, 25/4/2006.

scenario and 80 million tons under an optimistic scenario, while gas production could increase to 55–110 bcm per year by 2020: this huge variance of estimates suggests that these figures are closer to educated guesses than scientifically substantiated conclusions.[2]

Gazprom and the Siberian Division of the Academy of Sciences also provide radically different estimates of the maximum eastern oil and gas production (see Table 2).

Table 2: Maximum Oil and Gas Production Levels in East Siberia and the Far East

production	Gas in bcm		Oil in 10^6 tons	
	Gazprom	Academy	Gazprom	Academy
Sakhalin	72.2	30.0	23.1	35.0
Yakutiya	53.0	47.1	8.8	10.8
Irkutsk Region	46.3	53.0	9.1	13.4
Krasnoyarsk Krai	35.3	19.9	7.1	35.8
Total East Siberia and the Far East	206.8	150.0	48.1	95.0

Source: OGV, 17, 2005, p. 43.

The difference in estimates reflects the very low level of exploration of East Siberia and the Far East: 9% for onshore territories and 6% for the shelf, while some Arctic seas are practically unexplored. The average density of drilling in the region is 2 meters of deep wells per km^2, while in Russia on average it is 23 meters per km^2.[3]

Foreign companies operating in Russia, such as BP, are convinced that since all previous discoveries were made during Socialism, modern prospecting and exploration technologies will show that reserves are significantly larger. This view is confirmed by the recent

2 Energy Strategy of Russia up to 2020, Moscow, 2003.
3 Promyshlenniye Vedomosti, 20/5/2005.

discovery made by the Petromir company that in early 2007 booked a major Angaro-Lensk gas field in the Irkutsk Region.[4]

A purely geological issue of resource potential in the east of Russia also has political connotations, since competing governmental agencies (and in some cases certain companies) offer different forecasts of the future depending on their personal interests. The Ministry of Natural Resources (MNR), for example, holds quite optimistic views. Deputy Minister Anatoly Temkin anticipates production figures of 30 mln tons of oil and 50 bcm of gas in the immediate future, maintaining that available oil and gas reserves of the biggest fields can ensure these annual production levels.[5]

By contrast, railroad representatives are more skeptical: they doubt that oil production in East Siberia will increase from the current 0.5 mln tons to 30 mln tons by 2011. This skepticism is determined by a desire to promote their own version of oil deliveries to China and confidence in the advantages of railroad transportation over pipelines.[6]

There are currently four hydrocarbon production centers in the East of Russia: Irkutsk, Krasnoyarsk (including Taimyr and Evenkiya), Yakutsk, and Sakhalin. They all have substantial hydrocarbon reserves and have shown certain progress in the development of oil and gas fields. However, with the exception of the Sakhalin center, where commercial petroleum production is under way, all other centers mainly extract hydrocarbons for local needs. Being inadequately studied, these regions have a significant upside potential.[7]

Oil in East Siberia is of high quality, with a density of 0.87 gram/cubic meter, sulphur content of about 0.12%, and low paraffin content

4 It has gas reserves of 1.5 bcm in category C1 and 1.22 tcm in category C2; condensate reserves exist of 0.077 mln tons (category C1) and 61.79 mln tons (category C2).

5 Report of the Deputy Minister of Natural Resources A. Temkin, "On the Program of Geological Prospecting and Exploration of Oil and Gas Fields in East Siberia and the Far East of Russia", parliamentary hearings, the RF State Duma, 17/2/2005.

6 See, for example, Energy Vector of Russia's Eastern Geopolitics; one of the authors is V. Yakunin, head of Russian Railroads.

7 23.4% of the initial total reserves of Sakhalin have been explored, 18.3% of Irkutsk Region, 3.1% of Evenkiya, 2.6% of Taimyr, and 12.3% of Yakutiya (versus 26.2% for Russia on average). OGV, 11, 2004, p. 47.

(around 1%); it is superior to West Siberian crude and could be sold at a premium to the Urals blend.

The major gas fields of Krasnoyarsk Krai, the Irkutsk region, and Yakutiya radically differ from the fields of Western Siberia since they contain valuable components (ethane, propane, butane, helium) that can be used in the gas chemical industry. It is thus necessary to establish gas chemical enterprises and storage facilities for helium.

Most fields contain oil, gas, and condensate, which means that all components must be produced simultaneously and the construction of oil pipelines, gas pipelines, and condensate pipelines is required.

Oil and gas exploration and production in this Greenfield province will be much more expensive than in other parts of the country, largely because East Siberia and the Far East are characterized by harsh climatic conditions. The annual temperature spread of 70°–75°C in the Far East is one of the highest in the world. Nearly the whole territory of the Far East is covered by permafrost, which increases construction costs.

According to Academician Kontorovich, oil exploration will cost $ 2.40–2.60 per ton in West Siberia in the future. Gas exploration will cost less than $ 1 per 1000 m^3. However, in East Siberia and Far East onshore oil exploration will cost $ 3.70–4.50 per ton and gas up to $ 1.50 per 1000 m^3. On the Far Eastern shelf, oil exploration will cost $ 5.60 per ton and gas around $ 2 per 1000 m^3.[8]

TNK-BP believes that gas-lifting costs will be $ 28–35 per 1000 m^3 in the East of Russia, though in West Siberia Gazprom reports $ 3–4 and independent gas producers $ 2 per 1000 m^3.[9]

Enormous funds are needed to bring this new hydrocarbon province on line. According to the Ministry of Energy and Industry, to achieve the necessary volume of gas reserves (6.7 tcm) to ensure commercially viable gas production in the East of Russia by 2030, geological and exploration drilling of some 3.45 million meters will be necessary, which means an investment of at least $ 10 billion.[10]

8 OGV, 17, 2005, p. 35.

9 OGV, 14, p. 61.

10 Report of the Head of the Department of the State Energy Policy of the Ministry of Industry and Energy V. Saenko at the 5th All-Russian Oil and Gas Week, November 1, 2005.

According to Kontorovich, $ 87 billion are needed for the formation of major oil production centers in the East of Russia. To ensure crude production levels of 80 mt by 2030 in East Siberia and Yakutiya, some $ 14.5 billion will have to be invested in exploration, and $ 2.8–3 billion on the Sakhalin shelf.[11]

And when expenditures for creating the appropriate social infrastructure and general-purpose industrial facilities are factored in, the sums become truly awe-inspiring. Despite the windfall oil revenues that Russia enjoys today, there will hardly be enough domestic funds to develop this province.

The government intends to shift the main financial burden of developing the eastern oil and gas industry onto the companies. According to MNR, to ensure the necessary increment of reserves, $ 13–16 billion will be required by 2020; currently the federal budget outlays for the development of East Siberia are about $ 40 mln/yr, and by 2020 they will amount to a total of $ 800 mln – companies will thus have to invest the lion's share of funds.[12]

Though financially the eastern petroleum project might seem too challenging, it should not be regarded as a purely commercial endeavour since it has serious political connotations. The problems that the region faces are a major threat to Russia's economic and political security, and they stem both from the socialist legacy and the turbulent transition period.

11 OGV, 10, 2005, p. 15.

12 Report of the Deputy Minister of Natural Resources A. Temkin, "On the Program of Geological Prospecting and Exploration of Oil and Gas Fields in East Siberia and the Far East of Russia," parliamentary hearings, the RF State Duma, 17/2/2005.

2. Social and Economic Problems of East Siberia and the Far East

East Siberia and the Far East are depressed areas with the exception of natural resource sectors and export-oriented industries: the crisis of the 1990s was deeper and longer in the Far East than in the rest of Russia. Today the GDP in Russia is growing faster than in the Far East; this gap is constantly widening.

Since the late 1960s, mining industries have been developing faster in both East Siberia and the Far East than manufacturing sectors, and both regions have now turned into "resource colonies" for the rest of the country.

The population is very unevenly distributed there: it is mainly concentrated along the Trans-Siberian railway, which means that remote territories have not been developed. The average density of population is 2.1 persons/km^2 in East Siberia and 1.1 in the Far East. There is an acute shortage of manpower and, in particular, of qualified personnel.

The demographic problems of the East have become a serious threat to Russia's national security. In 1989–2002, the population of the Far East declined by 16%, from 8.0 to 6.7 million, through both natural attrition and out-migration, compared to a 4% decline in Russia as a whole. At the same time, illegal immigration from neighbouring countries is growing.

Despite salaries that are higher than the national average, the standard of living of the local population is lower than in other regions because all goods and services are more expensive. A huge disparity in economic development and social indicators in different eastern areas may result in serious discontent. Thus, the Far Eastern Federal District is characterized by vast differences between its constituent regions. The average salaries there in May 2007 (15,789 roubles) were somewhat higher than the average indicator for Russia (12,185 roubles), while the difference in average salaries between its regions was very large (30,773 roubles in Chukotka versus 11,372 in the Jewish Autonomous Republic).[13]

13 www.gks.ru.

Despite the significant fuel and energy potential of the Far East, over the 1990s the region became notorious for its energy crisis. The annual fuel deficit in the Far Eastern region is 7 mtoe which is covered by supplies from other regions of Russia.

Due to the growing centralization of power in Moscow in the recent past, the regions in East Siberia and the Far East have been deprived of many prerogatives and much autonomy, breeding dissatisfaction with the federal government.

The ability of the Eastern regions to resolve their problems largely depends on the lobbying capacity of regional administrations just as the ability of financial and industrial groups to promote their corporate interests to influence Moscow does, often to the detriment of the local economies, which also aggravates disparities among various regions.

Environmental problems are particularly serious in these regions: because of the prevailing low temperatures, the fragile ecosystems have a weaker recuperative capacity. Also, under socialism, many high-polluting industries were created there while no real attention was paid to environmental protection measures. The wasteful consumption of valuable natural resources (timber, fish, fur animals), which was always characteristic of the Soviet approach to this "colonial" part of the country, became practically barbaric during the transition period.

The limited development of the transportation infrastructure[14] and the absence of trunk pipelines in the East of Russia are the key obstacles to the creation of the new petroleum province. There are just two trunk oil pipelines from West Siberia to the Irkutsk Region, oil and gas pipelines from Okha (Sakhalin) to Komsomolsk-na-Amure, and gas distribution systems in Sakhalin and Yakutiya.

Oil companies would not be able to export the oil they could produce there; on the other hand, without sufficient volumes of oil to pump, the government was in no hurry to build a pipeline that would only incur losses. This situation has evolved into a vicious

14 In East Siberia, for example, the density of railroads is only 21 km per 10,000 km^2 versus 50 in Russia on average, and the density of car roads is 15 km per 10,000 km^2 versus 31 km in Russia on average. Therefore, winter roads (zimniks) and rivers are widely used for transportation purposes.

circle: oil was not produced in the absence of a pipeline, and a pipeline was not built because oil was not produced.

All the region's inherent problems worsened during the transition period. Economic ties between European Russia and this remote region were disrupted during the 1990s. Dmitry Trenin of the Carnegie Moscow Center believes that the current status of East Siberia and the Far East is the main domestic threat now faced by Russia. In the 21st century, the issue of national unity will be resolved in the east, which after the disintegration of the USSR suffered from depopulation, deindustrialization and general degradation. For this reason, it is vital for Russia to achieve "dual integration" of its eastern provinces by keeping them part of Russian territory while incorporating them into the dynamically growing APR markets.[15] Gennady Chufrin of IMEMO fears that Russia may lose its sovereignty over these regions, and not because of international intrigues or external conflicts.[16]

3. Moscow's Eastern Policy

During the 1990s, East Siberia and the Far East were largely forgotten by Moscow.

The *Energy Strategy of Russia up to 2020* stressed the need to develop eastern hydrocarbon resources; it further held that the share of Russian oil exports going to Asia-Pacific countries would grow from the current 3% to 30% in 2020, while that of natural gas would be 15%.[17] This document, however, mainly provides targets – without offering any real mechanisms for achieving them.

15 Vedomosti, 14/2/2005.
16 "Russia in APR", Russian Expert Review 5, 5/11/2003.
17 Energy Strategy of Russia up to 2020, Moscow, 2003. Today, 10–11 mln tons of oil and 7–8 mln tons of petroleum products are delivered to APR mainly by railroad from West Siberia and Sakhalin (less than 100,000 tons of oil are exported from Yakutiya and East Siberia). See: "Energy Vector of Russia's Eastern Geopolitics." Ekonomika, Moscow, 2006 p. 126.

Moreover, during the previous decade Russian oil and gas companies have largely focused their efforts on privatization, halting the production decline, and consolidating assets in other parts of the country; they regarded activities in this practically untapped area as economically unfeasible (particularly with low world oil prices).

However, the attitude of Moscow to its "eastern stepchild" seems to have changed in the new millennium. In contrast to Boris Yeltsin, Vladimir Putin frequently travelled to Siberia and the Far East, trying to understand their problems, and often visited key Asian countries to gain insight into this region.

Recently, Russia adopted a line of closer cooperation with China, India and other APR countries. The approach to the eastern areas of Russia changed accordingly. Putin formulated the goal of developing East Siberia and the Far East as one of the priorities for his second term in office; with high oil prices this project became more achievable financially.

In late 2006 the RF president assessed the situation in the Far East as "a threat to national security" and stressed the need "to invest money in the Far East". Consequently, the Ministry of Economic Development and Trade is developing the federal target program *The Development of the Far East and Zabaikal Region up to 2013*, which envisages investment of $ 14.3 billion in the territory.[18]

In practical terms, the Kremlin is implementing certain measures to transform these backward territories. First, the federal government is enlarging the federation subjects in the East: thus Taimyr, Evenkiya and Krasnoyarsk Krai were merged into a single unit in 2007, and the Ust-Ordynsk Autonomous District was added to the Irkutsk Region. Presumably, this is being done in order to strengthen Moscow's control over the regions.

Second, a major personnel reshuffle is underway. In line with the new policy of appointing governors, Putin replaced Boris Govorin, the governor of the Irkutsk Region, with Alexander Tishanin, and Konstantin Pulikovskiy, the authorized representative of the president in the Far Eastern Federal District, with Kamil Iskhakov, the former mayor of Kazan. Some new appointees, such as Alexei Kuzmitsky, the new governor of Kamchatka, have St. Petersburg

18 Center for Current Politics in Russia, 9/4/2007.

backgrounds.[19] Those who do not presumably have one thing in common: loyalty to the Kremlin in general and to the president in particular.

The third line of activities involves engaging big business in the solution of social and economic problems of the region. This trend was likely inspired by a successful experiment on Chukotka, where Roman Abramovich, after becoming governor in 2000, managed to radically improve the living standards of the local population. As a result, President Putin did not allow the oligarch to resign from this position in early 2007; he viewed Abramovich as a guarantor of social stability in Chukotka.[20] Consequently, the Far Eastern governors together with the relevant federal ministries are preparing to involve businessmen in revitalizing the Far East on the basis of the fashionable concept of state-private partnership.

The long-term efficiency of this state-private partnership is tested through an experiment that is underway in East Siberia. The program *Complex Development of the Lower Pre-Angara* became one of the four winners to receive financing from the state Investment Fund in June 2006. It envisages building and modernizing car roads, constructing a bridge across the Angara, building a railroad, completing the Boguchansk power station, and building an aluminium-producing plant and a pulp and paper enterprise. The construction of car roads would help to implement Rosneft's plans to develop the Yurubchen field (see below).

Simultaneously, the government is attempting to promote oil and gas development in the East of Russia in several ways:

Fiscal policy. The lack of appropriate tax incentives was a serious obstacle for companies that wished to work in the East of Russia. Now the government has evidently understood that investors in the region will require additional fiscal benefits for extracting hard-to-recover reserves. Changes in the Russian tax code concerning differentiation of the mineral production tax[21] are being considered.

19 Nezavisimaya gazeta, 1/6/2007.
20 Vedomosti, 12/2/2007.
21 This tax does not take into account mining, geological, and other specifics of individual fields, and its rate is too high for the development of low-profit fields.

These changes envisage a reduction of the rate for depleted fields, for the production of hard-to-recover reserves and tax holidays for new oil fields in territories that are not well explored, including East Siberia and the Far East. Possible preferential treatment for companies developing new fields on the shelf, mainly in the Arctic seas and the Far East, is being discussed. Gazprom and Rosneft, who divided the continental shelf among themselves, will benefit most from this fiscal innovation.[22]

Licensing policy. Currently, 54 zones with 900 mln tons of oil and 3.5 tcm of gas have been licensed off in the East of Russia.[23]

In 2005, MNR and the Federal Subsurface Use Agency developed the *Program of Licensing of East Siberia and Yakutiya.* It covers the territory of more than 600,000 km^2 and targets 213 promising zones that will be auctioned off. The priority of licensing is determined through 2015.[24]

The Law on the Subsurface. Efforts to modernize the 1992 Law on the Subsurface have been under way since 2004. Now, amendments to it are being discussed that are largely aimed at limiting the involvement of foreign majors in the development of Russian strategic deposits. By denying IOCs access to future auctions, they will ensure a lower price for key Russian companies such as Gazprom and Rosneft.

Presumably, MNR has a conflict of interest with the Federal Security Service: the latter considers the criteria for referring deposits to the category of strategic reserves as being too mild.[25] The ministry proposed designating oil and gas fields with booked reserves of more than 70 mln tons and more than 50 bcm, respectively, as strategic reserves. Since many fields in the East of Russia can be regarded as strategic, the new development of the Law on the Subsurface is particularly relevant for this region.

22 Vedomosti, 4/5/2007.

23 Report of the Deputy Minister of Natural Resources A. Temkin "On the Program of Geological Prospecting and Exploration of Oil and Gas Fields in East Siberia and the Far East of Russia," parliamentary hearings, the RF State Duma, 17/2/2005.

24 Oil of Russia, 4, 2005, p. 29.

25 Kommersant, 16/5/2007.

4. New Players and Their Key Projects

The desire of the Kremlin to establish total control over the strategic sectors of the economy (particularly oil and gas) and to reinforce its positions in the crucial eastern region will have the greatest impact on progress in developing the hydrocarbon potential of East Siberia and the Far East. Presumably, the government believes that the monumental task of revitalizing the East of Russia and forging energy ties with Asia can be entrusted only to companies under its control. To this end, Moscow is creating conditions for displacing private actors in this territory in favour of state-owned companies and for limiting the role of IOCs in favour of domestic entities (see Table 3).

Table 3: Key Eastern Players: 2000 and 2007

2000	2007
Russian companies	*Russian companies*
YUKOS Slavneft TNK SIDANCO Rosneft	Rosneft Gazprom Surgutneftegas TNK-BP
Foreign companies	*Foreign companies*
ExxonMobil ChevronTexaco BP RoyalDutch/Shell SODECO Mitsui Mitsubishi	ExxonMobil BP RoyalDutch/Shell SODECO Mitsui Mitsubishi Sinopec CNPC KNOC ONGC

Until recently, state companies were relatively poorly represented in the East of Russia, where private companies dominated the petroleum landscape. Gazprom had almost no presence in the region, while Rosneft, though having eastern assets and participating in important projects, was too weak to be considered a serious player. However, now Gazprom has significantly increased its eastern empire, and Rosneft has become the number one oil company in Russia. These two heavyweights grabbed the most important assets in the region; their success in implementing key projects and overcoming formidable eastern challenges and their conflicts of interests will largely determine the progress in the creation of the new eastern petroleum province, the relations between Russia and APR countries, and future directions of the eastbound hydrocarbon exports.

4.1 Gazprom

Over the last years the monopoly has created its eastern empire. First, wishing to strengthen its oil arm, it acquired 72.7% of Sibneft in 2005 and received assets of Sakhalin, Krasnoyarsk Krai and in Chukotka. Sibneft, currently Gazprom neft, owns the license for the Lopukhovo-Vostochno Shmidtovsky block with recoverable resources of 325 mtoe of Sakhalin-5.

Also, Sibneft's subsidiary, Sibneft-Chukotka, holds licenses valid until 2007 to develop three Chukotka oil and gas on-shore fields. In March 2002, Sibneft-Chukotka obtained a five-year geological license for three blocks of the Anadyr-1 project in the Bering Sea.

Gazprom neft also owns half of Slavneft. Slavneft is a relatively minor player with a production of 23.3 mln tons of oil in 2006. However, in the mid-1990s, Slavneft acquired licenses in Krasnoyarsk Krai for the Kuyumbinsk field acting on its own, and for the Tersko-Kamovsk block jointly with the Eastern Oil Company, a subsidiary of YUKOS.

The C1 and C2 reserves of the Kuyumbinsk and Tersko-Kamovsk zones are estimated to be 210 mln tons of oil and 174 bcm of gas. Slavneft produced 27,000 tons of oil from its fields in 2006. It expects to begin year-round development of the Kuyumbinsk field in 2007 and build a pipeline in 2009–2010. By 2011 the annual volume of oil production could be some 730,000 tons, and by 2012 it should reach 3 mln tons.[26]

Slavneft's position is rather peculiar: presently, Sibneft and TNK control about 99% of the company. New owners have divided the company's oil earmarked for export while leaving Slavneft a separate legal entity. Gazprom hopes to acquire Slavneft completely if it manages to get the half of Slavneft owned by TNK-BP.

Gazprom has taken another important step in strengthening its positions in the east of Russia by joining Sakhalin-2 as majority shareholder. Sakhalin-2 is based on Piltun-Astokhsk and Lunsk fields, with total reserves of 155 mln tons of oil and condensate and 490 bcm of gas.

On June 22, 1994, the first PSA in Russian history was signed. Sakhalin-2 was the only Sakhalin project without Russian involvement: until recently, the shareholders of Sakhalin Energy, the project operator, were Shell (55%), Mitsui (25%), and Mitsubishi (20%). The first stage of the project included development of the Piltun-Astokhsk field, which had been producing oil since 1999 from the Vityaz complex during the ice-free period. In May 2003, the second stage of the project began, focusing on development of the Lunsk gas condensate field. It envisages construction of:

- a 17 bcm/yr production platform on the Lunsk field;
- an onshore complex to prepare gas produced from the Lunsk field;
- two 800 km parallel oil and gas pipelines going from the north to the south of the island Sakhalin, to ice-free Aniva Bay;
- an oil export terminal;
- a 9.6 mt/yr LNG plant in Aniva Bay.[27]

Sakhalin-2 produced 1.6 mln tons of oil in 2006; these were delivered to Japan, South Korea, China, Taiwan, the Philippines, Thailand, and the US. Long-term agreements for Sakhalin LNG have been

26 See details at www.slavneft.ru.
27 See project details in www.sakhalinenergy.com.

signed with Japanese and Korean companies[28] as well as with Shell Eastern Trading Ltd (the latter envisaging deliveries of 1.7 mln tons of Russian LNG for 20 years to the US).

The progress of Sakhalin-2 has been far from smooth. As the project developed, protests of environmentalists became stronger. Under pressure from ecologists worried about the safety of gray whales, the consortium had to change the route of the sea pipeline and delay its construction.

But the project's real troubles started after Shell announced an increase in expenses and changed the schedule. The revised budget through 2014 was determined to be $ 20 billion, while the commencement of LNG deliveries was to be postponed from 2007 to 2008. These news displeased Vladmir Putin and Alexei Miller, CEO of Gazprom.

In September 2006, MNR revoked the positive environmental conclusion of the second stage of Sakhalin-2. The new environmental assessment could have taken up to one year while Sakhalin Energy had already signed contracts on LNG deliveries beginning in 2008, the disruption of which would have resulted in enormous losses. Sakhalin-2 faced financial problems since EBRD could not provide the promised credit without the environmental approval from the Russian authorities.

As a result, in December 2006, Gazprom acquired 50% plus one share in Sakhalin Energy for $ 7.45 billion. Each shareholder of Sakhalin Energy reduced its share by 50% with proportionate distribution of the payment. Experts assessed the controlling interest in Sakhalin-2 at $ 8 billion to $ 10 billion; Gazprom thus used environmental arguments to reduce this price.

Vladimir Putin congratulated project participants on a successful deal and promised that all their problems would be over.[29] Indeed, it is to be expected that all environmental considerations will be scrapped, gas exports will face no obstacles, and some compromise on the increased budget would be reached. Also, Gazprom's partners will probably get a chance to cooperate in other projects.

28 Korean Kogas, Japanese Tokyo Electric, Tokyo Gas, Kyushu Electric, Toho Gas, Hiroshima Gas, Tohoku Electric, Osaka Gas.

29 Vedomosti, 22/12/2006.

Later, in 2007, Gazprom finally obtained Kovykta in the Irkutsk Region. The C1+C2 category gas reserves of Kovykta were estimated to be 2.13 tcm in early 2002; gas condensate reserves were estimated at 108 mln tons. Kovykta's gas contains 0.28% helium.

Kovykta was discovered in 1987; at that time its reserves were assessed at 190 bcm, so it was regarded as a regional project and targeted for local gasification. To implement it, RUSIA Petroleum was established in 1992 by the administration of the Irkutsk Region, Angarsk Petrochemical Company, Irkutskenergo, and others.

Focused geological efforts began only in 1998 after Interros and BP got involved, and the real potential of Kovykta became clear in the late 1990s. The battles for Kovykta continued for a decade. Its development was delayed because of the complicated structure of RUSIA shareholding, which involved seriously conflicting interests. After the creation of the TNK-BP alliance in 2003, the company became the leader in RUSIA with 62.9%.

Initially, the project had a significant export orientation: in November 2003, RUSIA Petroleum, CNPC/PetroChina, and Kogas signed an agreement to deliver an international feasibility study. It was supposed that 4 bcm/yr would be provided to meet domestic needs. Some 30 bcm per year were to be exported to China (20 bcm/yr) and Korea (10 bcm/yr).[30] It was expected that the project would be approved by the governments of the three countries in March 2004.

These expectations were torpedoed by Gazprom. Gazprom's top priority is to preserve a single gas export channel and it insists that gas should be exported not from any given field, but from the USGS. Consequently, the monopoly leveraged its administrative resources to hinder the development of Kovykta, insisting that its gas should be delivered to the USGS for domestic consumers. It criticized the intention of RUSIA Petroleum to provide only 4 bcm of gas from Kovykta to the Irkutsk Region and to export 30 bcm. A study conducted by Gazprom's institutes showed that gas exports to China are fraught with price risks.

According to the licensing agreement, the commercial development of Kovykta was to commence in late 2006, while gas export was to begin in 2008. Without the construction of an export pipeline,

30 OGV, 18, 2005, pp. 92–94; Kommersant, 29/12/2005.

it would be impossible for gas production to reach the planned level of 34 bcm/yr. That means RUSIA would violate the licensing agreement. Hence, since 2003, MNR has been regularly threatening to revoke RUSIA's license.

In view of this, RUSIA decided to focus on the regional component of the project. In 2003, RUSIA Petroleum submitted its plan of gasification of the Irkutsk Region and in March 2004 established the East Siberian Gas Company (ESGC), together with the regional administration to carry out gasification.

However, following the RUSIA initiative, Gazprom signed a five-year framework agreement with the administration on gasification of the territory using small fields located in the north of the Irkutsk Region.

Since it became clear that it would be impossible to achieve anything without the consent of Gazprom, in March 2006 TNK-BP proposed to Gazprom the creation of the East Siberian Consortium; the concern would obtain 50% plus 1 share, and the rest would be owned by TNK-BP. This proposal was a compromise between RUSIA and Gazprom, because it could increase the share of gas used domestically.

Still, Gazprom claimed that Kovykta was not its top priority and the project proposed by RUSIA does not take into account certain important aspects, such as the full-scale extraction and use of helium.

It seems that Gazprom was playing a waiting game in order to get better terms from TNK-BP. Presumably a quick development of Kovykta does not meet the interests of Gazprom since this project will put an additional burden on the funds of the company that were to be focused on Yamal.

In September 2006, the procurator's office of the Irkutsk Region demanded suspension of the license for Kovykta development. The Irkutsk authorities claimed that RUSIA should have produced 9 bcm of gas by late 2006, but did not. Yet the region does not need that much gas; the license was issued in 1992, and since then, the region's gas demand decreased to 2.5 bcm. It may approach only 4 bcm by 2010, according to TNK-BP.[31]

31 Vedomosti, 24/1/2007.

MNR checked the fulfilment of licensing conditions by TNK-BP in January 2007, mainly focusing on gasification. In February, the commission of Rosnedra gave RUSIA three months to correct the discovered violations.

As a result of all these moves and counter-moves, in the summer of 2007 TNK-BP reached an agreement with Gazprom on selling its share of the project to the monopoly. Gazprom estimates the deal at $ 600–800 million, while TNK-BP hopes to get $ 700–900 million, presumably more or less the market price.

The new owner of RUSIA will have a much better chance of reaching an agreement on all debatable issues with the relevant officials. Gazprom is beginning to prepare a new plan of field development that will correspond to the Eastern Gas Program. Commercial gas production in Kovykta will begin in 2017, and the blue fuel will be delivered to the USGS to cover the potential shortage of gas in Russia (though a certain proportion of it may go to China).

Will Gazprom succeed in developing Kovykta? Success will hinge primarily on the results of negotiations with China, the leading potential market for Kovykta gas. To date, these negotiations are deadlocked because of price issues.[32] But the concern, as is its custom, represents this failure as a result of external factors, namely an agreement on Sakhalin-1 gas deliveries between ExxonMobil and CNPC.

Programs of gas industry development: Gazprom plays a crucial role in the region as coordinator of all gas activities and the exclusive exporter of gas. During the 1990s, Gazprom was not interested in the East of Russia; at that time Gazprom's had enough problems elsewhere. However, an interest emerged in the new millennium, and Gazprom is promoting its own vision of regional development.

In 2002, the Russian government appointed Gazprom coordinator for implementing the state's eastern gas policy and instructed Gazprom and the Ministry of Energy to develop a "program of creating a single system of gas production, transportation and supplies in East Siberia and the Far East with due account for possible

32 Vedomosti, 26/6/2007.

gas exports to China and other Asia-Pacific countries" (Eastern Gas Program).

The essence of Gazprom's strategy was to establish an eastern gas-producing center and to expand the USGS to East Siberia and the Far East. Two principle tasks were formulated; the priority of meeting domestic demand, and the implementation of an export policy on the basis of a single export channel.

The strategy was revised many times. An intermediate option envisaged three possible options of gas transportation ("west", "center" and "east").[33] Gazprom chose the "east" option[34] as the cheapest, with minimum export risks; this approach does not meet the objective of creating a single system of gas production and transportation in the region, however. This scenario envisaged conservation of the major Chayandinsk field[35] through 2030, thus hindering the economic development of Yakutiya.

Gazprom tried to justify its somewhat obstructive approach by high risks, which are increased by uncertainty in the Chinese market. Gazprom emphasized that China would need the main volumes of Russian pipeline gas only after 2015, and South Korea no earlier than in 2020.[36]

The development of the Eastern Gas Program was completed only in 2007. Even the latest version of the program contained 15 different scenarios of how to develop eastern hydrocarbon fields through 2030.[37]

33 See in greater detail Nina Poussenkova, "The Wild Wild East", Carnegie Working Paper, Moscow, 2007.

34 Four centers of gas production formed the basis of Gazprom's strategy: Irkutsk, Yakutsk, Krasnoyarsk and Sakhalin. Under the "east" option, the Krasnoyarsk and Irkutsk centers are to be connected to the USGS and supply domestic consumers of the region. The Yakutsk center is to work in the autonomous regime and to provide consumers of neighboring territories. Only Sakhalin gas is to be exported to Korea and China, both as LNG and through pipelines. Thus, two isolated systems are to be created, one to deliver East Siberian gas to the USGS and the other to ensure gas exports to Asia-Pacific countries.

35 Chayandinsk is Yakutiya's biggest field still held in the undistributed fund. Recoverable reserves of gas are 1.24 tcm, oil at 50 mt. Investments to develop the field are estimated to be $4.06 billion.

36 OGV, 17, 2005, p. 54.

37 Vedomosti, 23/3/2007.

Key disputes still concern the fate of the Kovykta and Chayandinsk fields. Gazprom believes that the East-50 plan, which holds that the east of the country will not need a single system of gas supplies until 2030, is the optimum scenario. The northern fields of the Irkutsk Region and the associated gas of the Krasnoyarsk Krai fields would meet the demand of the region, while the yield of Sakhalin-1 and Sakhalin-2 could be exported. Kovykta will be commissioned only in 2017. (However, the fate of Kovykta depends on Gazprom's agreement with TNK-BP.)

Gazprom expects that the program will be approved by the government by the end of 2007. The concern tried to use this occasion to drastically increase its own resource base. Gazprom officially declared that it intended to acquire the Chayandinsk field and blocks of Sakhalin-3 uncontested, for which changes in legislation will be required. It explains this proposal to bypass competitive bidding as the approach that best serves state interests. Under an agreement with Seoul, the concern is scheduled to deliver 10 bcm/yr to South Korea beginning in 2012–2013 via a Khabarovsk-Vladivostok pipeline which the concern will complete by 2011; according to Gazprom, however, "there is no resource base for this pipeline".[38]

Under the East-50 scenario, total investments in the eastern gas program to 2030 will be $ 60.1 bln (in 2006 prices). The program envisages gas production by 2010 at 27 bcm/yr, and 162 bcm/yr by 2030.[39] This result of a five-year effort leaves much to be desired and even the Ministry of Industry and Energy criticized it: a ministry spokesman has said that "the program does not contain anything destructive or anything that does not allow for an alternative solution," and admitted that the versions of the program were selected "under conditions of equal economic inefficiency."[40] It would probably be too optimistic to expect that such a document will help achieve a real breakthrough in gas industry development in the East of Russia.

38 Regional Economic Digest, 22/6/2007.
39 RBC Novosti, 26/6/2007.
40 Regional Economic Digest, 22/6/2007.

Relations with China: Currently, relations between Gazprom and China are expanding. In March 2006, Gazprom and CNPC signed a *Protocol on Deliveries of Natural Gas from Russia to China*. Deliveries are expected to begin in 2011, and in the future China will import up to 80 bcm of gas from Russia.

The 2006 agreement is an important step forward in implementing Russia's plans to diversify its gas export markets. However, it is unclear whether Gazprom will be able to provide enough gas to meet its long-term commitments to Europe, satisfy the growing domestic demand and ensure deliveries of gas to China.

Gazprom plans to pump gas to China from two directions, the *western direction* (Altai project) and the *eastern direction*. The latter is completely undeveloped.

The Altai pipeline is designed to pump 30 bcm/yr. Its advantage is the relatively short distance to the consumer and the fact that no transit countries are involved. To deliver gas from West Siberia, a new 2,800-km pipeline will be built in the existing Urengoi–Surgut–Chelyabinsk transportation corridor and then through the Tomsk region and Altai Krai to enter China going to the Sinytzyan-Uigursk Autonomous District in the west of China. There, it will be linked to the Chinese West–East pipeline, which will deliver gas to Shanghai.[41] The Altai project is based on the resources of the Nadym-Pur-Taz region, which may be insufficient. So there is a good chance that East Siberian gas will be needed to fill the pipeline. This plan is criticized by environmental NGOs since the pipeline is to pass through a national reserve area in Altai.

Thus, through aggressive asset grabbing and use of administrative resources, Gazprom has taken an important step towards implementing the concept of a single gas export channel in the East of Russia and restricting the activities of foreign majors. From a virtual player with high authority but no actual assets, it turned into a formidable force in the region.

41 See project details at www.gazprom.ru.

4.2 *Rosneft*

Having acquired Yuganskneftegas in 2004, Rosneft was transformed overnight from a minor player with 21 mln tons of oil production into the second largest oil company in Russia (85.6 mln tons of oil in 2006). After purchasing the lion's share of the remaining oil assets of YUKOS in spring 2007, Rosneft became the undisputed leader of the domestic oil industry; it now aspires to equality with ExxonMobil and BP by 2010. Russia's leadership probably wants to have a national oil company that is a match for the global supermajors, and Rosneft is being increasingly positioned as such an entity.[42]

Rosneft considers East Siberia and the Far East of Russia as the zone of its strategic interests. Rosneft has historically had a strong presence in the Far East: when the oil assets of Russia were divided in the early 1990s, it received Sakhalinmorneftegas (SMNG), which is involved in the Sakhalin-1, Sakhalin-3, -4 and -5 projects. Sergei Bogdanchikov, the current president of Rosneft, originally headed Sakhalinmorneftegas, and this region is psychologically important to him.

The positions of Rosneft are particularly strong on Sakhalin. The Sakhalin-1 project is based on the fields of Chaivo, Odoptu and Arkutun-Dagi, with total recoverable reserves of 307 mln tons of oil and condensate and 485 bcm of gas. Sakhalin-1 was the second project to be signed on PSA terms after Sakhalin-2. The presence among the shareholders of the state-owned Rosneft gave it additional weight. The shareholders of Sakhalin-1 are: Exxon Neftegas (30%), the project operator; Rosneft-Astra (8.5%); Sakhalinmorneftegas-Shelf (11.5%); SODECO (30%); and ONGC (20%).[43]

42 See in greater detail: N. Poussenkova, "Lord of the Rigs: Rosneft as a Mirror of Russia's Evolution. The Changing Role of National Oil Companies in International Energy Markets". J. Baker's Institute for Public Policy Publications, Houston, 2007.

43 In February 2001, the Indian ONGC joined the project getting its 20% from shares of Rosneft and Sakhalinmorneftegas.

The Sakhalin-1 project will consist of four stages:

First stage: the development of the Chaivo field. In 2005, Exxon began production from Chaivo with crude deliveries to the Komsomolsk refinery.

Oil production at the first stage will amount to 12.5 mln tons/yr which will be sold to domestic consumers and exported. For oil exports, a 221 km oil pipeline will be built to port De Kastri.

Gas production also began in 2005. 1.7 mcm of gas per day is supplied to Khabarovsk Krai.

Second stage: commissioning of the southern platform on Chaivo and gas exports by pipeline to Japan, when a gas pipeline is laid on the bottom of the Okhotsk Sea to Hokkaido Island. Nonetheless, there are doubts that this pipeline will be built because of Gazprom's position.

During the third and fourth stages, participants will maintain a production plateau and will produce oil from Arkutun-Dagi (after 2014) and develop gas reserves of Arkutun-Dagi and Odoptu (after 2032).[44]

The overall budget of the project was approved in 2003 and amounted to $ 12.8 billion in 2002 prices. The first stage of the project, worth $ 6.5 billion, has been completed. In 2005, ExxonMobil submitted an application to the Ministry of Energy and Industry to increase the budget to $ 17 billion.

Despite the credibility provided to Sakhalin-1 by the involvement of Rosneft, there were certain political problems connected with its implementation. In 2003, MNR criticized ExxonMobil for delaying geological and exploration works, and for a number of environmental violations. Presumably, these claims are consistent with the pressure now put on foreign companies operating in the Russian oil and gas sector.

A more serious problem is Gazprom's desire to maintain the single export channel and thus reach agreement with Rosneft concerning exports of Sakhalin-1 gas and the gasification of Khabarovsk Krai.

44 See project details at www.rosneft.ru and www.exxonmobil.com.

Currently, Rosneft is considering various directions for the gas exports from Sakhalin-1, including China, Japan and South Korea. Although Rosneft has the right to export Sakhalin-1 gas independent of Gazprom since the project is implemented under PSA terms, it admits that cooperation with the gas monopoly might be feasible.

Gazprom is vitally interested in Sakhalin-1 gas that could help it turn into a key player in the Far East. Therefore, when Exxon Neftegas signed a preliminary agreement with CNPC to build a 8 bcm/yr, 900-km pipeline from Sakhalin to China in October 2006, Gazprom strongly resisted the plan. Apparently, this agreement also displeased Vladimir Putin.[45]

In the summer of 2007, Alexander Ananenkov, deputy chairman of Gazprom, demanded that Sakhalin-1 gas should be delivered to the domestic market, as he estimated the gas demand of four Far Eastern regions to be 15 bcm/yr. "We consider it necessary to have a directive passed requiring Sakhalin-1 gas be sold to Gazprom for the gasification of Russian Federation subjects and not for export, as ExxonMobil wants," he unambiguously declared.[46]

The desire to get all the gas of Sakhalin-1 is not dictated by Gazprom's concern about Russian regions, but its desire to eliminate the competition of ExxonMobil, since the agreement with the Sakhalin-1 shareholders permits China to bid lower prices in negotiations with the concern.

Therefore, directions of gas deliveries from Sakhalin-1 will be determined by relations between Gazprom and Rosneft.

Sakhalin-3, Veninsky Block: In 2003, Rosneft obtained a five-year geological license for the Veninsky block within Sakhalin-3, with forecast resources of 169.35 mln tons oil and 258.07 bcm gas. In 2005, Rosneft signed an agreement with China's Sinopec to conduct exploration and prospecting of the region. Sinopec committed itself to taking on 25.5% of Rosneft's obligations during the exploration and development periods. Oil produced from the Veninsky block will be exported to China, Japan and Korea.

45 Vedomosti, 2/11/2006.
46 Vedomosti, 20/6/2007.

Sakhalin-4 and Sakhalin-5: An alliance of Rosneft (51%) and BP (49%) will conduct exploration on the Zapadno-Shmidtovsky block of Sakhalin-4 (oil and gas reserves of 181.2 mln tons and 281.14 bcm, respectively) jointly with the Kaigansko-Vasyukansky block (1,172.44 mln tons of forecast oil reserves), and the Vostochno-Shmidtovsky block (recoverable reserves of 411 mln tons oil and 255 bcm gas) of Sakhalin-5.

If exploration efforts are successful, Rosneft plans to develop all three blocks as a single project with a common infrastructure. The oil will be pumped through Rosneft's pipeline to De Kastri.

Rosneft intends to produce gas, provided its development and transportation are economically feasible - if Gazprom provides access to its export gas pipelines. BP agreed to take on part of the financing of all three projects.

In addition to Sakhalin, Rosneft has carved a niche for itself in East Siberia. First, it acquired the Vankor field located in Krasnoyarsk Krai. As of the late 1990s, its reserves were estimated at 105.1 million tons of light and sweet oil in the C2 category and 20.2 million tons in the C1 category; gas reserves were booked at 76.8 bcm. In 1993, JV Eniseineft won the license for Vankor. JV's founders were Eniseineftegasgeologiya and Eniseigeophysica, the administration of the Turukhansk district and the Anglo-Siberian Oil Company (ASOC), which was founded by, among others, two sheiks from the UAE.

The initial structure of Eniseineft's shareholders did not last long. First, YUKOS acquired 20% of Eniseineft. Then Total and Rosneft joined the game. In 2002, Total negotiated with the administration of Krasnoyarsk Krai, Eniseineft and ASOC to participate in field development. In February 2003, Total said it was ready to invest $ 3 billion in Vankor over two years. But its plans were torpedoed by Rosneft, which in 2003 bought ASOC for $ 80 million and became owner of a 59% stake in Eniseineft. The same year, Rosneft also won lawsuits against YUKOS.

Having gotten rid of its rivals, Rosneft started drilling at Vankor in March 2004. In March 2005 Rosneft announced that it had almost doubled the recoverable reserves of Vankor, bringing them to 250 mln tons. Commercial development was to begin in 2008. Peak oil production will amount to 19 mln tons/yr, and gas production to 5.1 bcm/yr. Project investments will total $ 6.5 billion.[47]

Rosneft's view concerning foreign involvement in the development of Vankor underwent a serious evolution as estimates of the field's reserves increased. In 2003, Sergei Bogdanchikov was saying that they would need a partner for Vankor. A year later, he stated that Rosneft would not think about engaging a partner until it received a full picture of reserves. By Summer 2006, when the reserve potential of Vankor was assessed, Bogdanchikov answered the question about inviting a foreigner to Vankor: "We'll never make such a gift to anybody. It is one of the most promising assets in the world, and we'll develop Vankor only by ourselves".[48]

Rosneft fought for Vankor not simply to expand its reserve base. By buying ASOC it did not permit a foreign major to get Vankor. If this had happened, a new channel of oil export not controlled by Transneft (read: the state) could have emerged.

Vankor oil could have been shipped along the Northern Sea Route, which would have avoided Transneft's bottlenecks and diversified markets: it could be delivered to Europe, America, and Asia. Also, it would have made sense not to mix high quality Vankor crude with the Urals blend in Transneft's system.

Rosneft studied several versions of oil transportation from Vankor. Finally, it decided to pump oil to ESPO, which experts viewed as a less attractive option than delivering crude by tankers along the Arctic seas. This decision was made by taking into account political rather than commercial considerations - after the government decided to build ESPO, for which Bogdanchikov lobbied, and which would be impossible to fill without Vankor. In some senses, the patronage of the federal authorities is a mixed blessing for the company, since it sometimes requires Rosneft to sacrifice profitability for political objectives.

Rosneft further strengthened its eastern positions further by buying 25.9% of Verkhnechonskneftegas, license holder for the Verkhnechonsk field, from Interros in 2005. The Verkhnechonsk field is located in the North of the Irkutsk Region and has C1+C2 reserves

47 See project details www.rosneft.ru.
48 Oil and Capital, 7, 2007 (special supplement), p. 5.

of 201.6 mln tons of light and sweet oil, 3.4 mln tons of condensate and 95.5 bcm of gas. In 1993, RUSIA Petroleum got a license for the field, but the company focused mainly on Kovykta and established Verkhnechonskneftegas in 2002 transferring the license to it.

In 2003, RUSIA sold shares of Verkhnechonskneftegas to its shareholders[49] for $ 1.2 million.

Verkhnechonskneftegas produced only 2,500 tons of oil in 2006 because of the absence of a transportation infrastructure. Peak oil production will amount to some 7.5 mln tons.

The involvement of now-mighty Rosneft could help with the project implementation since it is vitally interested in this oil to help fill the ESPO, which will now pass close to Verkhnechonsk.[50]

In the spring of 2007 Rosneft also acquired the remaining eastern assets of YUKOS, including its holdings in the Yurubcheno-Tokhomsk Zone, which are important for the company as resource base for ESPO

The Yurubcheno-Tokhomsk Zone (YTZ) located in Evenkiya is one of the most promising hydrocarbon areas of East Siberia. The total geological resources of the zone amount to 2.5–3.5 billion tons of liquid hydrocarbons and 0.5–1.2 tcm of gas.

In the mid-1990s, licenses for YTZ fields were auctioned off. The East-Siberian Oil and Gas Company, subsidiary of YUKOS, became the owner of the Yurubchen field (703 mln tons of recoverable oil resources); Slavneft and the Eastern Oil Company (where YUKOS had a controlling interest) also became co-owners of the Tersko-Kamovsk block.

Oil has already been produced from 75 exploratory wells in the Yurubchen field that yielded an average of about 100 tons/day (37,600 tons in 2003) of flowing oil compared to the national average of around 7 tons/day. Yurubchen oil is light and sweet; offsetting this, however, is the complex structure of the deposit with its extensive gas cap.

Yurubchen's development is further complicated by the total lack of a transportation infrastructure; therefore, the crude is sold to the local boiler houses.

49 BP, TNK, Interros, and the administration of the Irkutsk Region.
50 Oil and Capital, 9, 2006, pp. 9–12.

In 2001, YUKOS planned to achieve peak production (13 mln tons/yr) at the Yurubchen field within seven years. The investment requirements for the Yurubchen and Kuyumbinsk fields were estimated at $ 17 billion. Some $ 3 billion were to be spent on creating the infrastructure, primarily a pipeline and a road.

However, the progress achieved in developing the YTZ was undermined by the YUKOS case. In February 2005, oil production by the East-Siberian Oil and Gas Company was practically halted because YUKOS' shares were seized and, as a result, financing was suspended. Construction of the oil pipeline was also shelved.

Since the cessation of activities in the field threatened to disrupt the heating season in Evenkiya, in June 2005 Evenkiya authorities asked the Russian government to resolve this problem. In July 2005, German Gref, the Minister of Economic Development and Trade, proposed that Rosneft become the operator of the field.

In spring 2007, Rosneft bought YUKOS's assets in the YTZ; however it is to be expected that the change of ownership will further delay production activities within the zone.

Downstream. Rosneft's downstream positions in the East of Russia are also strong: initially, it owned the Komsomolsk refinery and two petroleum product distributors (Nakhodkanefteproduct and RN-Vostoknefteproduct). The company also has three export terminals in the Far East: the Vostochny Oil Terminal; De Kastri, with a capacity of 1.5 mln tons/yr; and Nakhodka, with a capacity of 6 mln tons/yr.

Rosneft is aggressively strengthening its downstream presence in the Far East. First, it is stepping up its activities at the Komsomolsk refinery with capacity of 6.4 mln tons/yr of crude and a refining depth of 60.1% (the bulk of crude from West Siberian fields comes by rail). In 1999, Rosneft began upgrading the refinery in order to improve the quality and enhance the yield of light products, particularly aviation kerosene and diesel fuel, which are in great demand in the Far East and Pacific Rim countries. (The refinery exports up to 80% of its products to Japan, South Korea and Vietnam.) Following the upgrading, production volume will amount to 7 mln tons/yr of petroleum products; the total investment in the upgrading will come to $ 700 million.

Rosneft acquired all the YUKOS refineries in spring 2007, including the Angarsk[51] and Achinsk[52] refineries in East Siberia and its fuel stations, including those located in East Siberia. Consequently, the company became number one in Russia in refining and marketing and inherited YUKOS' eastern positions; yet it remains to be seen whether Rosneft will be able to utilize them as efficiently as its predecessor did.

Also, Rosneft intends to build a 20 mln tons/yr refinery at the final point of ESPO by 2012 – in line with Russia's declared intentions of shifting from exports of crude to products with higher added value. The depth of the refining will amount to 93%. The refinery is expected to produce gasoline and diesel fuel corresponding to Euro-4 and Euro-5 standards, with 95% of output exported and 5% used for the needs of the Primorsk Krai.[53]

Then, Rosneft declared its interest in the Khabarovsk refinery,[54] which it would like to buy together with Alliance's network of fuel stations. The Alliance Group, that bought its controlling interest from SIDANCO, does not intend to sell these assets, but Rosneft does not hesitate to use its administrative clout to acquire companies it is interested in.

New geopolitical role. Rosneft now plays an important role in Russia's relations with China, South Korea and India because of the projects the company is undertaking in the east.

During Vladimir Putin's latest visit to Beijing, an agreement was signed between Rosneft and CNPC on cooperation in Russia and China. In mid-2006, Vostok Energy JV was established by Rosneft (51%) and CNPC (49%) to deal with hydrocarbon production and exploration in Russia. It appears that CNPC made a good business decision in choosing its Russian partner: in August 2007, Vostok Energy defeated Surgutneftegas and BASEL in an auction for two small fields in the Irkutsk Region that are located close to ESPO.[55]

51 The largest petrochemical company in the former USSR with a 19.18 mln tons/yr oil refinery, a chemical plant and storage facilities.
52 Capacity of 6.5 mln tons/yr and refining depth of 65.97%.
53 Vedomosti, 15/9/2006.
54 A capacity of 4.7 mln tons/yr and a refining depth of 61.4%.
55 Vedomosti, 1/8/2007.

Another Russian-Chinese JV will be established in China to deal with refining and marketing: a 10 mln ton refinery may be built there along with a network of 300–400 fuel stations.

Rosneft seems to have a special relationship with China, as the Chinese banks provided $ 6 billion for the Yuganskneftegas acquisition to be repaid by Rosneft's oil deliveries to China (48.8 mt of oil by 2010). Chinese companies have aspired to player status in the Russian oil industry for many years, but their achievements were practically nil before the Yugansk deal,[56] which seems to have changed their fortunes.

The Chinese further strengthened ties with Rosneft by becoming its shareholders. During Rosneft's IPO, CNPC bought shares valued at some $ 500 million.[57] Apparently, the Chinese were provided access to the Russian oil sector, and the question is whether Rosneft is the tail that wags the dog in Russian-Chinese relations.

The Chinese successes may be attributed to the strong socialist legacy of both Russia and China, which makes it easier for the Chinese to understand the specifics of doing business in Russia – particularly since they do not seem overly concerned about the issues of democracy.

Also, one key to the successes of Chinese companies in Russia is the fact that not only do try to access the Russian upstream, but they also let Rosneft in their downstream.

South Korea is another country with which Rosneft does business. In 2003, Rosneft got a geological license for the West Kamchatka shelf.[58] Since the project is risky and expensive, Rosneft in 2004 signed a memorandum of understanding with the Korea National Oil Corporation (KNOC). In 2005 KNOC acquired a 40-percent stake in the West Kamchatka Holding BV, with Rosneft retaining 60%.

56 The Chinese unsuccessfully tried to participate in Slavneft's privatization and in the Yuganskneftegas auction, and to buy the Orenbug company Stimul.

57 OGV, 13, 2006, p. 17.

58 The West Kamchatka petroleum region encompasses 105,000 km^2, including 80,000 km^{2}on the shelf, which is in the early stages of exploration. Total forecast resources are estimated at 1.03 bln tons of oil and 1.6 tcm of gas. Rosneft believes that this project is comparable in size to Sakhalin-1 and Sakhalin-2.

KNOC will incur all expenses connected with the geological exploration, as well as a significant portion of expenses during project implementation.

India, through ONGC, is another important partner of Rosneft. Their cooperation started in 2001 on Sakhalin-1. Then, in early 2007, Rosneft and ONGC signed a memorandum of understanding under which the Indians could get access to offshore fields, in return committing to pay for their development and admitting Rosneft to the Indian downstream.

Thus, both Gazprom and Rosneft managed to radically strengthen their positions in the east of Russia by limiting the involvement of foreign majors and displacing private domestic companies. "Etatisation" of the Russian oil and gas sector will probably continue in the future; however, it is unclear whether the economy of the country, the East of Russia and the population will benefit from this trend.

4.3 Rivalry between Gazprom and Rosneft

In the past, Rosneft was too weak to think about competing with the almighty Gazprom. Recently, however, it has emerged as its rival on different fronts, and it is winning in many instances. Rosneft got Yugansk although Gazprom was interested in it. Rosneft avoided integration into the gas giant; it bought Udmurtneft, a TNK-BP subsidiary, through Sinopec even though Gazprom was also a contender. Finally, Rosneft purchased the bulk of YUKOS in tough competition with the gas monopoly.

The rivalry between Rosneft and Gazprom is largely determined by contradictions between Dmitry Medvedev, the BoD chairman of Gazprom (and a "civilian"), and Igor Sechin, BoD chairman of Rosneft and leader of the "siloviki" group.

At present the competition between Gazprom and Rosneft is particularly fierce in the East. The two companies competed for subsurface plots in the Irkutsk Region, and Rosneft acquired the East-Sugdinsk block, having exceeded the starting price 25-fold and outbidding Gazprom.[59]

59 OGV, 26/12/2005.

A new trend in relationship of the two companies was established in November 2006 when they signed an agreement on strategic partnership through 2015. They intend to jointly produce, transport and process hydrocarbons, generate power, manufacture equipment, etc. Their cooperation will likely be limited, however, by the access Gazprom provides Rosneft to its pipelines. Since Gazprom insists upon maintaining its monopoly on gas exports, it might gradually cede the domestic gas market to independent gas producers (including Rosneft) who will be called upon to cover the potential domestic shortage of gas.

Some analysts maintain that this truce is a political decision. Vladimir Putin needs to mitigate the economic rivalry (at least) of the groups surrounding him on the eve of the 2008 elections; it is very unlikely that an across-the-board peace between the opponents can be forged within the foreseeable future.[60]

The rivalry between the adversaries will most likely intensify, as they will have some very tempting eastern assets to compete for.

There will be further disputes about the export of gas from Sakhalin-1 (see above) and about the Kirinsky block of Sakhalin-3.[61] Rumours are circulating in the market about the potential sale of the 50% stake in TNK-BP by its Russian shareholders – another sphere of dispute for the rivals. Lastly, Surgutneftegas may lose its independence and be devoured by either Rosneft or Gazprom.

This competition between Gazprom and Rosneft aggravates the current instability in Russia's oil and gas sector (and does not contribute to efficient development of the east of Russia). Tension in the upper echelons of power in Russia will only become more acute with the approaching elections of 2008.[62] From an optimistic point of view, however, it is better to have two dominating companies in the domestic energy sector than one; they ensure a system of checks

60 Oil and Capital, 12, 2006, p. 7.

61 In 1993, Mobil and Texaco won the tender for the Kirinsky block (700 bcm of gas and 70 mln tons of condensate) to be developed under PSA terms. In 1997 Mobil and Texaco were persuaded to concede one-third of their stakes to Rosneft and SMNG. In 2004, the RF government denied ExxonMobil and Chevron the right to work within Sakhalin-3/Kirinskiy block. Now this block is to be auctioned off again.

62 Oil and Capital, 8, 2006, p. 4.

and balances whereas a single company would encounter no barriers in its expansion drive.

5. Eastern Challenges: ESPO

On December 31, 2004, Prime Minister Mikhail Fradkov signed an ordinance on the construction of the East Siberia–Pacific Ocean (ESPO) pipeline: Taishet–Skovorodino–Perevoznaya Bay.

Initially, ESPO envisaged construction of the 1,020-mm pipeline with a total length of 4,188 km and a throughput capacity of 80 mln tons/yr of oil for the Taishet-Skovorodino section and 50 mln tons/yr for the Skovorodino-Pacific Ocean section. The resource base of the first stage was to be provided by the fields of West Siberia, as well as several East Siberian fields that are close to the route of the future pipeline. However, ESPO faced unexpected obstacles - the "ecological weapon" was successfully tested in East Siberia.

The initial plan for the ESPO route passed 800 m from Lake Baikal in a highly seismic zone (up to 10° on the Richter scale). This option brought numerous complaints from environmental NGOs. The State Environmental Expert Assessment Commission twice rejected it. However, Transneft pushed through its version of the route, claiming that all necessary safety measures would be taken. Transneft explained that moving the pipeline to the north was too difficult since it would pass through mountainous areas without inhabitants and infrastructure. Semyon Vainstock, president of Transneft, claimed protest demonstrations against the pipeline were intrigues of either foreign "puppeteers" afraid of Russia's stronger influence or domestic business structures that wished to protect their corporate interests.[63] Finally, in April 2006, Vladimir Putin ordered the pipeline moved 400 km to the north of Baikal.

Now, the pipeline's end point will be in Kozmino Bay and it will pass close to the oil fields of Yakutiya and the Irkutsk Region. The decision certainly was to the benefit of Surgutneftegas (the license

63 Vedomosti, 25/1/2007.

holder of the Talakan field), Verkhnechonskneftegas, the Irkutsk Oil Company, and Urals Energy,[64] since they, together with Rosneft, will provide oil to ESPO. As a result, some experts claimed that the protests of environmentalists were organized by some business groups. In any case, it is clear that environmental issues in Russia will be more actively used in the future to put pressure on certain companies. Sadly enough, though the threat to Baikal arising from the construction of the pipeline was eliminated, the lake is still in danger from transportation of crude oil by railway.

ESPO will be implemented in two stages:

1) By the end of 2008, Transneft will build the Taishet–Skovorodino pipeline with a total capacity of 30 mln tons/yr and a terminal in the Kozmino Bay to accept crude from Skovorodino arriving by rail.
2) Afterwards, Transneft plans to build an oil pipeline from Skovorodino to the terminal. This pipeline will have a total capacity of 50 mln tons/yr, bringing the throughput capacity of the oil pipeline system to Skovorodino to 80 mln tons/yr, of which 30 mln tons/yr is planned for transport to China.[65] The date of completion for the second stage has not yet been determined - according to Transneft, "everything will depend on the pace of geological and exploration works and development of the fields undertaken by the oilmen. We will build the second stage by the time they will be able to guarantee the necessary volumes of oil".[66]

Andrei Dementiev, Deputy Minister of Industry and Energy, nostalgically compared ESPO to BAM.[67] Indeed, these two projects are comparable in scale. However, BAM was the stillborn child of

64 Talakan is located in Yakutiya near the Chayandinsk field. It has recoverable reserves of 124 mln tons of oil and 47 bcm of gas, and geological resources of 300 mln tons. Surgutneftegas extracted 242,800 tons in 2006 and targets 3 mln tons by 2008, and 10 mln tons/yr by 2010. It is expected that by 2010 the Irkutsk Oil Company will be able to provide 1 mln tons of oil for ESPO. Urals Energy plans to produce 590 thousand tons by 2008, and 1.4 mln tons by 2011 from the Dulisminsk field in the Irkutsk Region.

65 See Energy Vector of Russia's Eastern Geopolitics, p. 72–80.

66 Vedomosti, 20/6/2007.

67 Vedomosti, 13/7/2007.

socialism. Hopefully, fate will be kinder to the pipeline child of the state capitalism; still its implementation faces many challenges:

Resource base. West Siberia will provide the pipeline with 30 mln tons/yr in the foreseeable future. Some 7 to 10 mln tons of East Siberian oil will be supplied to ESPO in 2009. Afterwards, volumes of East Siberian oil should reach 20 mln tons in 2010 and 40–45 mln tons by 2015. However, the reserves of East Siberia are still relatively small.

Rosneft, Gazprom neft and LUKOIL already deliver oil to China by rail guarantee oil deliveries via ESPO from West Siberia. Rosneft, which is in charge of ensuring supplies of oil to ESPO, promised that it would be able to fill 30% of capacity by 2009 and 50% by 2014.[68]

Upgrading the existing pipelines. Transneft is refurbishing the trunk pipelines from West Siberia.[69] By the time ESPO is commissioned, the pipeline upgrades should help increase the overall capacity of the oil pipeline system to the west of Taishet to 55 mln tons/yr. This amount exceeds the eastbound volume expected to be achieved by the time ESPO's first phase is launched. Presumably, Transneft intends to create excess capacity in case there is a delay in commissioning the East Siberian fields and West Siberia is called on to fill the gap.

Competition with Russian Railways. Another uncertainty is connected to ESPO's competition with the Russian Railways (RZD), which have always played an important role in transporting crude and petroleum products in the east of Russia and which plan to increase the total shipping capacity on the China-bound route to 30 mln tons/yr of oil. The competition between RZD and Transneft will be largely decided by the lobbying potential of the organizations. The railway lobby is quite powerful in Russia; besides, the RF government wishes to revitalize BAM, which can be used to transport certain amounts of oil.

68 Vedomosti, 20/6/2007.

69 Omsk-Anzhero-Sudzhensk – Angarsk/Irkutsk pipeline was commissioned in 1964; its length is 2,334 km, pipe diameter is 720 mm and design capacity is 12 mln tons/yr. Anzhero-Sudzhensk – Krasnoyarsk – Angarsk/Irkutsk pipeline was commissioned in 1975; its length is 1,441 km, pipe diameter is 1,020 mm and design capacity is 32 mln tons/yr.

Pipeline tariffs. The question of pipeline tariffs is also still open. Initially, the tariff for pumping oil through ESPO was to be $ 38.80/ton for the Taishet–Skovorodino section. For West Siberian oil, Transneft proposed $ 49.80 per ton. With this tariff, transportation costs to deliver oil from East Siberia to the Pacific Ocean would be more than twice as high as the transportation along the existing routes are. The higher transportation costs will be partly offset by the premium for the East Siberian oil in the APR markets as compared to Europe, which mainly receives the Urals blend. The additional 400 km involved in rerouting the pipeline and increased project costs will result in even higher tariffs. According to German Greff, to make exports through ESPO profitable, the highest pumping tariff should not exceed $ 50/ton; in this case the pay-back period for the first stage of the pipeline will be ten years.[70]

Financing. Initially, the project cost was projected to be $ 11.5 billion (including $ 6.65 billion for the first stage). Given the extended pipeline route, however, just the first stage will cost $ 11 billion. According to Semyon Vainstock, it is difficult now to forecast the cost of the second stage because of the rapidly growing prices for pipes.[71] If the decision on ESPO construction were made today, the decision-makers would have second thoughts, comments Valery Nesterov.[72] On the other hand, the decision on ESPO construction was made taking into account political considerations, which would probably have prevailed over commercial feasibility in any case. A branch of the pipeline to China from Skovorodino will be built using a $ 400 million grant to be provided by CNPC.

* * *

The development of oil and gas potential of the East of Russia and a breakthrough to the APR energy markets thus face formidable challenges:

70 Vedomosti, 19/7/2007.
71 RIA Novosti, 12/7/2007.
72 Vedomosti, 20/7/2007.

- the lack of a long-term strategic vision from the Russian;
- adverse climatic and complicated geological conditions of Russia's east;
- the acute social, economic and environmental problems of the region;
- demographic problems, including shortages of labor;
- the lack of a transportation infrastructure;
- enormous investment requirements;
- unclear hydrocarbon resources potential of the region;
- uncertainty with respect to ownership rights, exacerbated by the current redistribution of assets in the oil sector of Russia;
- the replacement of private companies with state entities (Rosneft and Gazprom);
- the rivalry between Rosneft and Gazprom in the region;
- Gazprom's policy of safeguarding its exclusive right to export gas and its desire to control all major gas projects;
- the financial difficulties of Rosneft after its recent acquisitions;
- worsening relations with international majors who can provide the necessary funds, technology and know-how;
- complicated relations in the Russia–China–Japan triangle;
- uncertainty with respect to the Chinese market, and unresolved gas pricing issues; and finally
- the intense competition with other suppliers in APR markets.

Fyodor Lukyanov

Russia–EU Relations: The Impact of the Energy Boom

It is a well known fact that oil and gas export play an extremely important role in Russian foreign policy. The export of hydrocarbons is considered by Russian officials as an advantage of the country in international relations and simultaneously as a strategic resource to strengthen Russia's influence in the global policy. During the last years the energy agenda has been turning the score in Russia–EU relations. Despite their mutual economic dependence, both sides have unsuccessfully attempted to impose their own strategy and vision of cooperation in this sphere upon the other. The future of Russia–EU relations in energy issues causes many questions and different assessments. The goal of this article is to examine different factors related to Russian oil and gas export that have influenced Russia–EU relations during the energy boom and try to predict what effect energy issues will have upon these relations in the closest future.

In mid-2003, crude oil prices on global markets exceeded the average world price in the period from 1947 to 2007 ($ 26.16 per barrel). Since then, oil prices – and the demand for oil – have been quickly and steadily growing.[1]

As opposed to previous peak periods – in 1974 after the Arab oil embargo, and in 1980 after the Islamic Revolution in Iran – it is reliability of supply, rather than oil prices per se, that is now on top of the agenda for the developed world. The global economy is now much more adaptive to price hikes than 30 years ago. No wonder that efforts to ensure energy security have become the core of the foreign policies of major oil-consuming countries. At the same time, new opportunities, both economic and political, have opened for oil producing countries. Russia, which is the world's largest hydrocarbon supplier, could remain on the sidelines of this process just as

1 http://www.wtrg.com/prices.htm.

little as the European Union, the largest and longest-standing client of Russia's oil and gas sector.

On the one hand, the importance of the situation on the global oil markets for Russia's behaviour on the international stage is very high. At the same time, it should not be overrated because the riches that have befallen Russia thanks to the oil price boom have not changed the essence of Russian policy and the aspirations and notions inherent in it. Yet, the "oil lever" has become a strong stimulant and has given a boost to the process of Russia's new self-assertion in the world. Most likely, this process would have taken place in a different economic situation on the oil and gas markets, as well; but now it has become rapid. Russian leaders have received a chance to quickly and efficiently implement ideas that otherwise would have taken much more effort and time.

The year 2003 was marked by crucial changes in Russia's policy. Inside the country, the authorities began to take ever more companies that ensured oil income for their owners under state control (the YUKOS affair offers the best-knows example of that). Problems that had seemed insoluble even a year before now appeared to be insignificant (for example, the so-called "Problem 2003". The year 2003 was critical in terms of debt repayment, as Russia had to pay the bulk of its foreign debts – 17 billion dollars). The very function of the state changed. Immediately after he came to power, Vladimir Putin spoke much about the need to ensure economic freedom and protect businesses against excessive state interference.[2] The state desperately needed money then and hoped for private initiative. Putin's statements during his second term of office demonstrate his increased propensity toward dirigisme and universal state control, as the growing inflow of petrodollars needs to be controlled and distributed.

Symbolically, the first really tough statement by Putin about the European Union was made in 2003:

2 See, for example, the president's July 8, 2000 Address to the Federal Assembly http://www.kremlin.ru/appears/2000/07/08/0000_type63372type63374type82634_28782.shtml (in Russian).

> The position of EU bureaucrats is wrong and unfair. We view this position as "arm twisting". But Russia's "arms" are getting increasingly stronger, and even such strong partners as those in the European Union cannot twist them."[3]

Putin referred to negotiations on Russia's accession to the World Trade Organization, namely to Brussels' demand that Russia raise gas prices at home to the world level. Soon, however, such an approach became universal in Russia–EU relations. Putin's predecessor left him an unenviable legacy in relations with Europe. The last years of Boris Yeltsin's presidency were marked by a profound disappointment of the West with Russia. Financial and corruption scandals, inconsistency in politics, the Second Chechen War, and the appointment by Yeltsin of a KGB officer as his successor degraded Russia's relations with the EU to a very low point. However, the framework of Russia–EU relations and the model of interaction remained unchanged.

This model was based on the Partnership and Cooperation Agreement, signed in the summer of 1994 and ratified by Russia in the autumn of 1997. Without analyzing this document, we can point out its main idea, namely the Europeanization of Russia. Without claiming a right to membership in the European Union (an idea that only inveterate idealists voiced even then), Russia would nevertheless gradually approach the "European model." The adoption by Russia of norms and rules of EU legislation, which was recognized by default as more progressive and better than Russia's, was intended to serve as an instrument of this rapprochement.

Scholars may now argue at length over whether or not that idea of integration was realistic. What is more important is that both parties embraced it. For the European Union, such an approach was quite natural. Those years were marked by an active integration of Central and East European countries, and it seemed that the model of pro-European transformation was good for the entire post-Communist world. However, Russia's readiness to abide by EU criteria was rather due to its revolutionary inertia. Democrats that came to power in Russia following the break-up of the Soviet Union at first had no doubts as to which way the country should go.

3 http://www.kremlin.ru/text/publications/2003/10/53889.shtml (in Russian).

They waged their struggle against the Communist system with the same slogans that their counterparts in Eastern Europe and other former Soviet republics used: "Down with Communism!", "Down with the Empire!" and "Forward to the West!" Soon, it turned out, however, that what was a goal for East European democrats was a means for democrats in Russia. As for the goal that the huge former empire was to seek, it remained unclear. Therefore the ruling class continued to believe that the country should follow the path taken by its former allies in the Communist bloc.

During his first years in office, Vladimir Putin followed the same path, although the upheavals of the late 1990s delivered a serious blow to Russia's "European choice." In the autumn of 1999, Putin, who was then the prime minister, was introduced to the European establishment at an informal EU summit in Helsinki. The spring of 2000, despite the fierce criticism by the EU of the developments in Chechnya, saw the beginning of an obvious warming of Russia–EU relations. By the end of the first year of Putin's presidency, a major initiative was announced in the energy field. The EU–Russia summit, held in November 2000 in Paris, discussed plans to build up Russian energy supplies to Europe (unofficially, it was described as the Prodi Plan). Russia and the EU established an energy dialogue. Putin's Address to the Federal Assembly in April 2001 said: "Integration with Europe is one of the key areas of our foreign policy."[4]

During his visit to Germany in September 2001, two weeks after the tragedies in New York and Washington, Vladimir Putin made a serious statement:

> If regional conflicts arise somewhere, Russia is ready to supply additional amounts of hydrocarbons to compensate for a possible drop in supplies to the European market. I am saying that in earnest.[5]

The same attitude dominated at the EU–Russia summit in Brussels in October 2001, at which the decision was reached to establish a High Level Working Group for elaborating the concept of a Common European Economic Space.

4 http://www.kremlin.ru/eng/speeches/2001/04/03/0000_type70029type82912_70660.shtml.

5 http://www.kremlin.ru/eng/speeches/2001/09/26/0001_type82914type84779_136380.shtml.

The period from the autumn of 2001 to the spring of 2002 was, perhaps, the culmination of the declarative rapprochement, a kind of "honeymoon" between Russia and the European Union. Yet, the very first practical contact on the issue of the transit of Russian citizens to and from the Kaliningrad Region after the accession of Poland and Lithuania to the EU caused a tense conflict between the parties. At the Moscow summit in May 2002, Putin actually accused the European Union of violating human rights:

> We are offered solutions that will actually mean only one thing, namely that the right of a Russian citizen to free communication with his or her relatives living in one or another part of the country will depend on a foreign state's decision.[6]

A year later, in his Address to the Federal Assembly, Putin reiterated that integration into Europe was Russia's historical choice; yet he added that the interests of "Greater Europe ... require that serious steps be made towards each other."[7] The key words here are "towards each other." It is the need for reciprocity, rather than unilateral moves, that would soon become the basis of Russia's European policy.

Putin expressed the same idea more graphically in May 2006, after the Russia–EU summit in Sochi:

> Full implementation of the Energy Charter and the additional provisions on transit would mean that we provide free access to oil and gas production and to the transport infrastructure. Of course we cannot help but ask what we are to get in return ... So we are not against the idea of doing this in the future, but we need to know what we would get in return. This is very easy to understand if you just think back to childhood when you go into the street with a sweet in your hand and another kid says, "Give it to me". And you clutch your little fist tight around it and say, "And what do I get then?".[8]

The period between 2002 and 2005 was the time of transition from declarative relations between Russia and the European Union to practical relations. No wonder that very serious friction between

6 http://www.kremlin.ru/appears/2002/05/29/0001_type63377_30092.shtml (in Russian).

7 http://www.kremlin.ru/eng/speeches/2003/05/16/0000_type70029type82912_44692.shtml.

8 http://www.kremlin.ru/eng/speeches/2006/05/25/2359_type82915_106123.shtml.

the two parties immediately emerged, as their interests objectively differed in many issues. The first disputed issue was Kaliningrad; then there were long and painful negotiations on the extension of the PCA to the new EU member states (2004); and finally there were endless delays with the signing of WTO accession protocols between Russia and the European Union (2005).

All those developments proceeded amidst the growing oil prices and changes in the post-Soviet space. Simultaneously, the Russian political system grew increasingly centralized. This process, which marked the final departure from the European model, was directly related to changes on the energy market. There is a very popular theory nowadays in the West about confrontation between two types of capitalism – liberal capitalism (personified by the United States, Europe and their allies) and authoritarian capitalism (Russia and China).[9] This division, which emerged quite recently, now is universally recognized. Even a recent *Freedom House* report speaks of the "Russian" and "Chinese" models, which even enrich each other.[10]

Uniting Russia and China into one model of "authoritarian capitalism" is a strange idea. In the case of China, it is still possible to say that the authoritarian political system is the source of economic growth. In Russia, the situation is opposite: authoritarianism is not the source but a product of growth, which initially was stimulated by fast-growing energy prices. As mentioned above, the strong inflow of petrodollars changed the very function of the state in Russia.

After 2003–2004, the state focused its efforts upon taking control over the more profitable sectors of the economy and the distribution of petrodollars. That inevitably caused the strengthening of the state machinery in all spheres – in the economy, and in home and foreign policies. The tendency toward monopoly climaxed by the end of Vladimir Putin's presidency. In politics, the choice was

9 See, for example, Azar Gat, "The Return of Authoritarian Great Powers", Foreign Affairs, July–August, 2007, pp. 59–69.

10 Countries at the Crossroads 2007 Overview Essay "Ruling By Law: Ambitions and Limits of the 21st Century Authoritarian Model" http://www.freedomhouse.org/template.cfm?page=140&edition=8&ccrpage=42.

finally made for one political party (the president's decision to head the electoral list for the United Russia party). In the economy, the emphasis was made on the establishment of large state monopolies in all sectors of the economy.

Interestingly, both President Putin and the main ideologist of today's Russia, Vladislav Surkov, Deputy Chief of Staff of the Russian president, continue to speak about Russia's European essence:

> Above all else, Russia was, is, and will, of course, be a major European power. Achieved through much suffering by European culture, the ideals of freedom, human rights, justice and democracy have for many centuries been our society's determining values. For three centuries, we – together with the other European nations – passed hand in hand through the reforms of the Enlightenment, the difficulties of emerging parliamentarism, municipal and judiciary branches, and the establishment of similar legal systems … We did this together, sometimes behind and sometimes ahead of European standards.[11]

> Our difficulties with the West are translation problems, problems of communication between related yet profoundly different European cultures … Our long-standing neighborhood and the vaguely felt distant kinship in religion only increase irritation. Differences between akin people, as you know, are more emotionally charged than between strangers or people indifferent to each other.[12]

The Kremlin no longer uses the old tradition, deeply rooted in Russian culture, of seeing Russia and the West as opposed to each other. Instead, Moscow demands an "equal" and "worthy" place at the western "table." Russian conservatives often criticize Europe for abandoning its own legacy and its Christian roots, which, on the contrary, are being restored and strengthened in Russia. Natalya Narochnitskaya, a historian and deputy of the State Duma, stated the following:

> Even the most uneducated person instinctively feels that he does not want to be part of the present degenerated Europe … If you look at their European Constitution, it is a most dull specimen of a liberal Gosplan [State Planning

11 http://www.kremlin.ru/eng/speeches/2005/04/25/2031_type70029type82912_87086.shtml.

12 V. Surkov. "Russkaya Politicheskaya Kultura – Vzglyad iz Utopii" [Russian Political Culture – The View from Utopia] [in Russian], http://www.edinros.ru/news.html?id=121456.

> Committee]. There is not a single value in it, for which great Europeans of the past mounted the scaffold.[13]

Generally speaking, the financial prosperity of the Russian state has largely increased messianic tunes in the political discourse.

Returning to Russia's relations with the EU, by the year 2005 it became clear that the model based on Russia–EU rapprochement and embodied in the PCA had already outlived its usefulness. Timofei Bordachev and Arkady Moshes reflected the following:

> There is something schizophrenic about the Russian–European relations, because neither party wishes to openly admit that they represent absolutely different political and economic systems. Therefore, their integration is unfeasible, at least in the mid-term. And if there is no chance for Russia's membership in the EU, why should Moscow adopt its political and legal standards?[14]

The period from the summer of 2005 to the autumn of 2006 was the most interesting time. At that time, Russia and the European Union launched an experiment designed to put aside the values issue, a matter of heated debates over a long time, and start building their mutual relations on the basis of interests only - especially because the complementarity and interdependence of their economies were obvious.

Actually, Russia proposed a new interpretation of the word "integration." The former meaning (harmonization of legislations, and economic, political and legal practices) gave way to "asset swap." In his famous speech in Munich in February 2007, Vladimir Putin explained this approach in the following way:

> I shall remind you of the transaction that took place between Gazprom and BASF. As a matter of fact, this was an asset swap. We are ready to continue to work this way. We are ready. But in each concrete instance we must understand what we give, what our partners give, calculate, have an independent international expert evaluation, and then make a decision ... We have

13 http://www.narochnitskaia.ru/cgi-bin/main.cgi?item=1r400r070306184751. See also the address of Patriarch Alexy II of Moscow and All Russia to the Parliamentary Assembly of the Council of Europe http://www.coe.int/t/dc/files/pa_session/sept_2007/20071002_disc_patriarche_en.asp.

14 T. Bordachev and A. Moshes, "Is the Europeanization of Russia Over?", Russia in Global Affairs, April–June, 2004, http://eng.globalaffairs.ru/numbers/7/526.html.

> actually just recently done something similar with our Italian partners, with the company ENI.[15]

Actually, such an approach once underlay the European integration, as the European Coal and Steel Community was also a kind of asset swap. However, Russia and the EU lack a key element for real rapprochement such as the one that took place in Europe during the 1950s. Before concluding the Montan Union, its participants (France, Germany, Italy and the Benelux countries) reached agreement on the general understanding of what a modern state was and how it should function. It was this understanding that was transformed into those common values.

Russia and the EU lack such an understanding. This means that they are unable to agree on any interests, as they speak different languages. Even when they use the same words, they mean different things. It is this principled mutual misunderstanding – and not Polish meat, not even the murders of Anna Politkovskaya and Alexander Litvinenko – that is the real cause of the absolute deadlock in Russia–EU relations. In the autumn of 2006, the two parties were still full of enthusiasm about their negotiations on a new basic agreement in place of the PCA. A year later, these negotiations were no longer mentioned. The EU's hopes to convince Russia to play according to European rules and – in particular – to ratify the Energy Charter, did not come true, and Moscow's attempt to achieve an important agreement with Europe – namely the consistent unification of their energy infrastructures on the principles of mutual benefit – was also not successful.

The years 2008 and 2009 will be crucial for Russia–EU relations. The following factors will be of major importance:

- First, the settlement of the issue of power in Russia. The stability of the established political model and the Kremlin's ability to pursue an aggressive foreign policy, as it did during the last year of Putin's presidency, will depend on the new political configuration.
- Second, Russia's ability to solve the growing economic problems at home. The inefficiency of the economic system is beginning to

15 http://www.kremlin.ru/eng/speeches/2007/02/10/0138_type82912type82914type82917type84779_118123.shtml.

show even when oil prices are high. Under certain circumstances, high prices will even aggravate domestic disproportions and the absence of structural reforms.

- Third, the EU's overcoming the institutional deadlock, in which it has found itself after the failure of the European Constitution. The deep crisis of the political model and the common European identity of the EU is a major factor in the deadlock in its relations with Russia.
- Fourth, the EU's ability to build a common energy policy, at least in relations with its outside partners. In this respect, the September 2007 proposals of the European Commission,[16] which introduce the principle of mutual market opening into the energy relations between the EU and its outside partners, may become the basis for a new start in the energy dialogue with Russia. Formally, this dialogue has been underway since 2000; actually, it has long been deadlocked.

Russia–EU relations are, perhaps, the best mirror reflecting the transformation of post-Soviet Russia – from the illusions of the 1990s to disillusionment in the early 2000s to euphoria of 2007. Both partners are now in a state of difficult transition that is far from complete. The most important thing is that the final destination remains unknown. Contrary to tradition, it is not only Russia that is an unpredictable partner today. The future of the European Union is also obscure, just as the entire situation in the world. Today, deep interdependence in the energy sphere is not so much a guarantee of successful development as a source of irritation and apprehensions for both parties. However, the economic reality and development in the world will inevitably push Russia and the EU towards each other again. And the new rapprochement will be based on a foundation with more solidity than the rosy yet rather groundless hopes of the early 2000s.

16 The EU Electricity & Gas Markets: Third Legislative Package, September 2007 http://ec.europa.eu/energy/electricity/package_2007/index_en.htm.

Alexej Yablokov

The De-Ecologization of Russia: Causes and Effects

The present review is not a detailed list of all environmental problems in Russia but a mosaic showing that environmental problems are one of the most disturbing issues. However, in almost all cases, they receive insufficient attention by the authorities.

A more detailed description of all Russian environmental problems as well as the references to concrete sources of given figures and facts can be found in A. Yablokov's report "Russia: health of nature and people" (in the series "Ecological policy", published by "APPLE". Moscow, 2007, 224 pages; in the Internet http://www.rus-green.ru).

1. Introduction

In Russia, according to the Constitution and a number of laws, the ecological information, information about environmental conditions, people's health, ecological and other threats cannot be kept secret. However, a number of people - not only experts but ordinary citizens as well - are pointing out with increasing alarm that the ecological information that is reported in Russia has become less detailed and unimportant. Instead of the analysis of tendencies in the development of oncological and other ecology-dependent diseases, the mass-media are filled with messages about fires and stings.

It is difficult to make correct administrative decisions without the knowledge of the true state of affairs. If one agrees with nuclear scientists that "it is time to forget Chernobyl", then it is possible to sharply reduce the support of the affected territories and liquidators, but if it is known that the 20-year-old latent period for cancers is now coming to an end, then it is necessary to become seriously concerned with the preparation for the inevitable surge of cancer in all 16 constituent territories of the Russian Federation which were under the clouds of Chernobyl in 1986. If it is known that in ecologically unsuccessful territories life expectancy is 5–8 years shorter than in others, then every responsible politician must to work out measures thar first single out and thereafter remove such dangerous situations, as well as determining how to make people's lives in these territories easier. Unfortunately, the Russian government often operates as if it were unaware of environmental problems and acts according to the Stalin-Goebbels principle: "no information – no problem".

Today in Russia new pollution is being added to old Soviet burdens. This process has sharply intensified after the liquidation of the state structures on the protection of the environment in 2000. Nature protection laws were enforced 12–15 years ago. Today the watering down of laws and the breakdown of state management in this sphere have increased the pollution problem. Even in schools "ecology" has not been an obligatory subject since 2001. The actions of the government are defined by the ideology of de-ecologization, which can be formulated in this way: it is necessary to use natural resources as much as possible; as ecological laws and rules make things complicated, it is necessary to change them; it is necessary to suppress ecological information so that the society does not worry about environmental problems; and finally, in order to carry all this out, it is necessary to sharply weaken the state's ecological structures and use different methods to keep public ecological organisations quiet.

It is significant that federal budgetary expenditures on the preservation of the environment, which in 1995–2003 accounted for less than 0.5% of the federal budget, have been reduced since 2007 – to less than 0.1%.

In the pursuit of unceasing enrichment Russia has apparently entered a new period of the Middle Ages:

- an uncountable quantity of illegal dumps has arisen everywhere around settlements;
- emissions into the atmosphere are cleared out completely in no region – there are regions where emissions are not cleared at all;

- sewage has been cleared out completely in only one region – again, there are regions where sewage does not get cleared at all.

In all developed countries economic growth has been accompanied by an improvement of the environmental quality and an increase in average life expectancy. In Russia, where the average salary has increased by a factor of 10 during the last 15 years, average life expectancy has not only not increased but has even decreased in some places! The average life expectancy of men in Russia today is approximately 59 years and that of women is about 72 years. In 2004, according to the world organization of public health services, 492,000 deaths in Russia were caused by ecological factors. This is 12–14 times more than death due to accidents and 15–16 times more than death from drunkenness.

It is necessary to take into the consideration the fact that the official data on which the present review is based are incomplete and that the system of the state ecological monitoring and control is not sufficiently representative. Here there are four examples:

- The number of posts in the Russian meteorology center for the supervision of the quality of superficial waters has been reduced several times during the last years;
- in some regions more than 200,000 km^2 fall under the control of one environmental inspector (by comparison, the area of the Great Britain is 244,000 km^2);
- over 90% of accidents in the oil industry are concealed;
- medical statistics including data on the infant death rate are distorted.

Huge territories remain outside a supervision network and a considerable portions of infringements upon nature protection rules and norms remain undiscovered. It is necessary to add that many enterprise that produce pollutants either do not submit data on the emissions and dumps or represent them in a deformed way, and state statistics on the use of natural resources (mineral, timber, fishing, etc.) as well as on diseases (and even the infant mortality rate!), need a considerable expert adjustment.

2. Population Health and Environmental Contamination

The sickness rate and the death rate of inhabitants of strongly polluted industrial centers (such as, for example, Dzerzhinsk, Irkutsk, Kemerovo, Krasnoyarsk, Novokuznetsk, Norilsk, and Cherepovets) are 1.5–3 times higher than the country average. In towns the atmosphere pollution caused by motor transport is the cause of almost half of the respiratory diseases. In the polluted territories the number of people sick with chronic bronchitis and bronchial asthma is steadily growing. The contribution of the atmosphere pollution in towns to overall sickness exceeds 40%.

A high level of lead content in the organisms of children has been documented near metallurgic and storage manufacturies in the North Ossetia, Kursk, Chelyabinsk, Sverdlovsk, and Saratov regions as well as in the territories upon which several thousand tons (!) of lead were thrown from helicopters in an attempt to extinguish radioactive fires caused by emissions that evaporated from the burning reactor in Chernobyl. In the meantime 2 million children in cities of Russia have behavioral and educational problems caused by the high lead content of the environment.

A high level of dioxin pollution – and connected with it, diseases in the population – can be observed in towns such as Tchapaevsk, Ufa, Volgograd, Novomoskovsk, Yaroslavl, Dzerzhinsk, Serpukhov, Novocheboksarsk, Slavgorod, and Usolye-Siberian.

The occurrence of malignant new growths among children is between 2 and 9 times the Russian average in Moscow, Saratov, Amur, the Chita regions, the Altai Republic, and the Khanty-Mansi autonomous area. The number of newborns with development anomalies is growing in the Novosibirsk area, the Smolensk and Penza regions, Kalmykia, Chuvashiya, the Arkhangelsk region, Primorye Territory, Tomsk, Sakhalin, and the Samara regions.

There is no doubt that environmental contamination (both now and in the past) is an important reason that the mortality rate in Russia is the highest in Europe.

In Moscow approximately 11,000 people died of air pollution in 2005 – three times the number caused by motor transport. It is not an exaggeration to state that ecologically-dependent sicknesses affect most Russian people on an annual basis.

3. Air Pollution

Since 2000 one can observe the growth of industrial emissions into the atmosphere. Leaders in the industrial pollution of the atmosphere are the Krasnoyarsk region, the Khanty-Mansi autonomous area, Sverdlovsk, Kemerovo, and the Chelyabinsk regions. Since 1990 a large contribution to air pollution has been made by forest fires (smoke from forest fires in Siberia reaches an area equal to that of the Great Lakes in Northern America).

61.5 million people in Russia live in conditions of an extremely polluted atmosphere and this number is growing every year. Table 1 shows the relative population rate (in percent) in different constituent territories of the Russian Federation living on the territories with "high" (H) and "very high" (VH) air pollution level.

Every second Russian experiences the negative influence of the elevated concentration of weighed particles (dust) in the atmosphere, and more than 2.4 million people are exposed to the influence of a concentration of more than 300 mkg/m^3 (for comparison: in the USA where the population rate is twice as high and industry is far more developed, a little more than 2 million people are exposed to this influence).

Table 1: The Relative Number (%) of the Urban Population in the Constituent Territories of the Russian Federation in Territories with "High" (H) and "Very High" (VH) Air Pollution Levels in 2003–2004

The constituent territories of the Russian Federation	% of citizens living under the conditions of H and VH air pollution level
Tajmyrsky atonomy	100
Moscow	100
St.-Petersburg	96.0–100
Samara area	86.2
Omsk area	78.4
Orenburg area	78.4
Kamchatka area	74.6
Tuva	73.4
Irkutsk area	67.0–73.1
Novosibirsk area	72.8–79.0
Khabarovsk territory	72.7
Astrakhan area	72.5

4. Water Pollution

36.1% of all sewage dumped into reservoirs and on the relief is polluted (including 7.4% crude). Huge anthropogenous pollution has led to the condition that only 12–14% of lakes and rivers in Russia are ecologically healthy.

The pollution of Lake Baikal – caused by air fallout, illegal buildings in the water protection zone, and by the drains of the Baikal pulp-and-paper combine – keeps growing.

The quality of underground waters is worsening everywhere. In 2003–2004 only 1% of surface reservoirs (with the existing level of water preparation) had potable water meeting hygienic requirements. Every second inhabitant of the Russian Federation uses potable water that does not meet these requirements.

The main reasons for the catastrophic quality of potable water in Russia are the pollution of surface and underground reservoirs, the absence or inadequate condition of their sanitary protection zones, the absence or unsatisfactory work of treatment facilities and disinfecting installations, and an overaged network of treatment plants, 40–70% of which require replacement.

5. Earth and Soil Pollution

In the main industrial and agricultural areas of Russia, the majority of lands had already reached an ecologically unsatisfactory condition in the 1950s to 1970s. From 1989 to 2007 the process of soil fertility loss for the reasons of erosion, infringements of land tenure, reduction of the quantity of fertilizers used, and chemical and radiation pollution accelerated. About 40% of farmlands are exposed to wind erosion and about 18% to water erosion. On the average, more than 11% of the territories of inhabited settlements (intended for building zones) across the country have been dangerously polluted by heavy metal compounds. In the Tomsk and Murmansk regions and Khabarovsk area, such territories occupy more than half of all inhabited places.

An unacceptably high level of sanitary and hygienic (microbiological, parasitic) pollution of territories (including inhabited settlements) is a natural consequence of the neglect of sanitary cleaning of inhabited places from production wastes and consumption by state bodies (including the universal occurrence of unauthorised dumps); an unsatisfactory quality or the absence of a centralised water drain in a number of large settlements; an unsatisfactory condition of sewer networks: all these serve as indicators of the inadmissibly scornful attitude of the government towards people's health.

Safe storage of firm industrial and household waste (FHW) – the quantity is constantly growing – remains one of the biggest unresolved problems. In Kaluga, Kursk, Lipetsk, Murmansk, Tambov, the Chelyabinsk regions, the Altay area, and the Koryak autonomous area, all landfills of FHW do not meet sanitary norms.

Across the whole country there are hundreds of thousands of unauthorized dumps (in the territory and region of Moscow alone 3000 exist). No more than 5 to 6% of secondary raw materials are used, compared to up to 80% in developed countries. Dangerous waste from other countries is also dumped on the territory of Russia.

47,700 km^2 of Altay territory, Yakutia, Khakassia, Tuva, Komi and also of Arkhangelsk, Kamchatka, Murmansk, Novosibirsk, Omsk, Tomsk, Tyumen areas and Yamal-Nenets autonomous area are polluted with 10,000 tons of metallic rocket constructions and remains of toxic rocket fuel components from the realization of space programs.

On the territory of the majority of constituent territories of the Russian Federation, there are ill-conditioned warehouses and burial placed for storing forbidden and obsolete pesticides and agrochemicals. 10,000 tons of poison gases and chemical ammunition were destroyed in many places in the 1970s and 1980s (flooded, blown up, burnt, dug) using methods that are ecologically dangerous.

During the last 15–20 years, areas of city territories affected by flooding and landslides have increased by 50–60%. The flooding of territories is a result of intensive "point" building when the foundations of skyscrapers dug deeply into the ground disturb the prevailing hydro-geological conditions.

The lack of attention to the preservation of the earth is leading to a growth of ravines and loss of fertile soils that is more rapid than in the last century. In southern regions (the Stavropol and Krasnodar areas, Kalmykia, Astrakhan, and the Rostov regions) the process of desertification is developing with incredible speed.

6. The Problem of Radiation Pollution

On average the level of medical irradiation (0.97 mzv per year) across the country has closely approached the permissible maximum (1 mzv per year).

Huge territories were polluted in 1949–1963 by radioactive atmospheric fallouts from nuclear weapon tests. As a result of several radiation accidents in the Southern Ural Mountains, large territories of the Chelyabinsk, Orenburg and Sverdlovsk regions became radiation polluted. During the 81st "peaceful" nuclear explosion on the area of 16 constituent territories of the Russian Federation, the atmospheric emissions of radionuclide apparently totaled a few million curie. Those radionuclides that remained in the bowels of the earth became the centers of constant (for many decades!) radiation pollution of underground and subsoil waters.

A big danger comes from radionuclide pollution of underground water pools by the enterprises of Severska (Tomsk region), Zheleznogorsk (Krasnoyarsk area), Dimitrovgrad (Ulyanovsk region), and "Beacon" (Chelyabinsk region). Long-living radionuclide pollution from Zheleznogorsk is traced across Yenisei on 1,500 kilometres up to the Arctic ocean.

The dangerous level of radioactive pollution of territories that resulted after the explosion of the 4th block of the Chernobyl atomic power station (1986) has remained up to the present on more than 1.2% of the territory of the European part of Russia (56,000 km^2). About 200,000 people now live in the polluted territories of the Bryansk, Tula, Kaluga, and Oryol regions.

Station storehouses at all Russian atomic power stations are filled to 60–80% with spent nuclear fuel, the storage and processing of which are connected with unresolved environmental problems. At the same time, since 2000 Russian borders have been opened for spent nuclear fuel from other countries – this will actually transform Russia into the radioactive dump of the world. The changes of the legislation accepted in 2000 (only the parliamentary fraction "APPLE" unanimously voted against this) make this threat real.

Hundreds of lost, ownerless radio isotope termo-electrogenerators (from more than 1000) that were used in the 1990s as independent power supplies for beacons and meteorological stations now represent a real threat of radiation pollution to the coasts and shelves of the Baltic, Barentsevo, Arctic and Far East seas.

7. The Problem of Oil Pollution

Huge territories of Western Siberia and the European North are polluted with floods of oil and chisel slime. The burning of a billion cubic metres of the passing gas continues (though in smaller scales than in the late 1980s) and each year destroys more oxygen than all our large cities consume.

Hundred million tons of oil-slime, a million cubic metres of black oiled water and a billion tons of ground polluted by oil and oil products have accumulated on the territory of Russia. By estimations, about 10 million tons of oil are lost annually during extraction and transportation. Experts estimate that about 30% of the poured oil (4–5 million tons) finds its way into reservoirs. Therefore even such large rivers as the Ob, Volga, and the Northen Dvina have mineral oil and derivate levels that are hundreds of times higher than before the beginning of oil extraction; they exceed safe levels by considerable amounts. The annual amounts of mineral oil carried by some rivers are up to 600,000 t for the river Ob, up to 360,000 t for the Yenisei, up to 82,000 t for the Volga, and up to 50,000 t for the Lena.

By expert estimations about 50,000 cases of pipeline breaks happen annually in Russia. On the whole, it is possible that not less than 5% of Russian territory is dangerously polluted by mineral oil. Among those that suffer most from oil pollution are Western Siberia, the North Caucasus, the Republic Komis, Bashkortostan, Tatarstan, and also Northern Prekaspiy and the Average Volga region.

8. The State of the Ecosystem, Vegetation, and Fauna

In the 20th century, steppes, forest-steppes, and deciduous woods of the European part of Russia as ecosystems have appeared to be on the verge of disappearance, mountain woods in the south are degrading, and the area of inundated landscapes in the pools of many large rivers has been reduced. The ecosystems of practically all big rivers of the Caspian, Azov, Black and Baltic seas and also the

Ob and Yenisei are irreversibly deformed; the rivers themselves have spread into cascades of weak stream water basins with extensive flooding of adjoining territories. Dams have created insuperable barriers to the migration of sturgeons, salmons and other valuable fishes. The building of flat hydroelectric power stations from the 1940s until the 1980s has made huge areas of the most fertile soil unusable and has destroyed ecosystems and the natural balance in the pools of the Volga, Don, Ob, Yenisei, Lena, and Amur. More than 3000 water basins have pressure head constructions that are in a dangerous technical condition.

The ecosystem of the Azov Sea – once the largest fish-producing sea in the world – is completely destroyed.

According to the official statistics, about 100 million m^3 of wood were felled annually in 1999–2005. Illegal woodcutting is a huge problem; in some areas of the Far East and the North it surpasses legal cutting by many times.

A number of species of hunting-trade animals are in danger of extinction, including the saiga, argali, and the polar bear. The number of the sturgeons in the Volgo-Caspian pool has been reduced by 90% in the last 25 years and trade resources in the Barentsevo and the Far East seas have essentially decreased. During the last decade the illegal use of bio resources has taken on an industrial scale, and in fishery it surpasses the official craft.

529 kinds of plants and mushrooms as well as 414 species and subspecies of animals are included in the red book of the Russian Federation – a state list of rare and vanishing species of animals and plants which are under protection. The number of some rare species of animals in nature is dangerously low: for example, there are only 300–400 Ussuriisk tigers left and 35–40 leopards. On the average, in Russia 1–2 species of fauna disappear each year and each 4–5 years 1 species of flora disappears.

A desperate situation has developed within especially protected natural territories. While in 1992–2000 26 new reserves and 18 national parks were created in the Russian Federation, from 2001 to 2005 no new reserves were created. The state of the existing reserved territories is under threat: the government tolerates and promotes the individual use of and (illegal) building in protected territories.

9. What Can Be Done?

The state ecological policy of Russia suggested by the party "APPLE" is based on the urgent necessity of transition from modern de-ecologization of the country to the ecologization of the governmental management, the industry and other branches of the economy, the whole economy and education.

Ecological trouble has become a brake on the social and economic development and a real threat for the steady political development of Russia. The environmental problems of Russia shout: it is impossible to live in such environment!

The requirement of the ecologization of the internal and foreign policy of Russia is one of the most important stances of the pre-election platform of the party "APPLE". If the Russian voter knows the truth about the ecological situation in the country, his or her region, about the dangerous influence of ecological factors on health, he or she will support the party's political program.

The highest priorities in the ecological policy of Russia are the following:

- the restoration of a specially authorized federal body (independent from the bodies on the use of natural resources) for the preservation of the environment;
- the restoration of the uniform system of state ecological examination;
- the expansion and strengthening of the system of the nature protection prosecutor's office;
- the restoration of ecological budgetary funds where all means for pollution of the environment should be concentrated;
- the removal of illegally erected constructions in water security and other nature protection zones (in cases of the self-capture of the earth without compensation, in case of the legal acquisition with indemnity to owners and criminal prosecution of the officials who have broken the nature protection legislation);
- the establishment of a system of information to the population concerning the quality of the environment and the linkage of

the environmental state with a person's health (on the basis of the restoration and the development of the state system of monitoring);

- the restoration of the nature protection legislation that was cancelled and watered down in 2000–2007 in full scale in order to close loopholes in the legislation allowing few people to enrich themselves at the expense of national natural resources;
- the restoration of an obligatory course on ecology in the secondary education system.

Natalia Zubarevich

Regional Development and Big Business in Russia's Regions

Regional development and big business are two different topics but in Russia they are closely linked with each other. The reason for that is that the dominating influence of first nature factors (raw natural resources) persists and the role of second nature factors – ones that have an agglomeration effect (economics of scale) and use human capital – are not adequately developed. This classification of factors, proposed as the New Economic Geography,[1] helps us understand the role of big business in the process of regional development. The production and refining of mineral resources is controlled by large companies, both state and private. The taxes paid by these companies make up the bulk of the regional budget surplus; additionally the companies, especially oil and gas companies, are often the main investors of the region. Besides tax money, business usually ends up financing social functions and programs in the regions. The social responsibility of these businesses is especially high in monofunctional towns where the company's main resources are located. Thus we see that big business plays a huge role in the economic and social development of the regions.

In this article the author tries to answer the following questions:
- How are Russia's regions and cities developing and how does big business influences this process?
- How did the geography of big business change – what are the zones of expansion?
- How do relations between big business and the regions' administrative organs evolve?
- What possibilities do these relations between big business and regional administrations have under the new conditions of

1 P. R. Krugman, Geography and Trade. 1999: MIT Press, Cambridge, MA.

de-privatization (the overtaking of private companies by state companies) and the globalization of Russian business at the same time?

1. The Tendency of Territorial Development

The Polarization of Regional Development

A stable tendency of the post-Soviet era is the concentration of Russia's economy in regions that have special benefits; the country's capital, Moscow, and the leading oil and gas producers in the Tumenskaya Region (together with autonomous regions). These regions produce more than a third of the Gross Regional Product (GRP) of the country as a whole (Table 1). Moscow's main income comes not from the service industry but from the income taxes of the big natural resource companies whose headquarters are located there. In the last few years the economic contribution from the regions Moscow and St. Petersburg also became significant. The economic growth of the second capital is due not only to special benefits but mostly thanks to the economics of scale with its agglomeration effect. Another significant factor are the administrative measures of the central government to relocate the headquarters of several corporations that transport oil and refine gas and are under government control to St. Petersburg. In all other leader regions big business plays a decisive role in the economy, either controlling the property needed for producing or refining natural resources or financing the spectre of services for diversifying investments (Moscow region, Krasnodar region). In result the share of the 10 leading economic areas grew from 46 to 57% of the national economy in 10 years.

Industrial dynamics also show an increasing polarization of the regions (Table 2). The leader positions have been firmly held by regions producing oil gas and metals – the most important export-oriented industries of Russia's economy – since the middle of the 1990s. In the last years the development of large agglomerations in the refining industry has accelerated. This is a positive tendency for

Table 1: The Biggest Regions' Share of Russia's Gross Regional Product in Percent

Leaders	1996	2000	2003	2006
Moscow	12.2	21.0	21.1	23.1
Tyumen region	9.4	9.9	10.3	11.7
Moscow region	3.6	3.2	3.9	4.2
St. Petersburg	3.4	3.3	3.8	3.6
Sverdlovsk region	3.5	2.7	2.7	2.9
Tatarstan Republic	3.0	3.3	2.8	2.7
Krasnoyarsk region	3.0	3.5	2.4	2.6
Bashkortostan Republic	2.9	2.6	2.4	2.3
Samara region	3.1	2.5	2.4	2.2
Krasnodar region	2.3	2.4	2.4	2.2
Total 10 regions	46.4	54.4	54.2	57.5

Table 2: The Industrial Output Index 2007 Compared to 1990 in Percent (1990 = 100%)

Leaders in %	2007	Outsiders in %	2007
Nenets autonomous district*	377	Khabarovsky region (Far East)	44
Leningrad region	198	Kamchatsky region (Far East)	46
Sakhalin region*	184	Rep. Nothern Osetia (South)	45
Belgorod region	164	Magadan region (Far East)	43
Moscow region	133	Kurgan region (Urals)	40
Rostov region	123	Rep. Altay (Siberia)	40
Tatarstan Republic*	123	Amur region (Far East)	37
Orenburg region*	111	Chita region (Siberia)	36
Tyumen region*	103	Rep. Kalmykia (South)	22
		Rep. Ingushetia (South)	12
Russia total	74		

* oil and gas extracting regions.

the country's raw resource based economy. Slowly developing and depressive regions in the south and east of the country are falling ever further behind the leaders.

Investments in Russia have three geographic directions -oil and gas producing regions, metallurgy regions, and large agglomerations of Federal cities (Moscow and St. Petersburg) (Table 3). The last years have seen investment in coastal regions of Southern Russia, especially in the Krasnodar Region where the 2014 Olympic Games will take place. Here, also, the main investors are large resource companies – metallurgy and energy. The outsiders in terms of investments are the depressive regions, the southern republics and most of the eastern regions, with the exception of those producing oil. This investment geography as a result increases economic disproportions between the regions and contibutes to the backwardness of the eastern regions.

In conclusion, we can point out three groups of leaders of economic growth in Russian regions. They are very similar to leader regions in other developing countries because their growth is based on natural benefits:

- large agglomerations with a growing service sector (the Moscow and St. Petersburg agglomerations)
- regions that produce export resources (oil, gas) and do basic refining;
- seaside regions that lie on the main routes of international trade (the Leningrad and Kaliningrad regions in the west and Krasnodar region in the south).

No regions with high-tech refining industries are among the leaders. This is a result of the domination of first nature factors – the abundance of natural resources and convenient geography, while the influence of second nature factors such as human capital is low when it comes to regional development.

The growth of regional inequality is not only present in economics but the same goes for the most important social components. The Gini index for regional inequality calculated by us shows a number of social indicators and especially the unemployment rate because the condition of regional labour markets is closely linked

Table 3: Per Capita Investment* (2002–2006) for Different Groups of Regions Compared to the Russian Average in Percent (Russia = 100%)

Leaders	to Russia average,%	Outsiders	to Russia average,%
Oil and gas extracting regions		*Depressive regions*	
Nenets auton. district	1619	Pskov region	44
Yamal-Nenets auton. district	989	Ulyanovsk region	44
Khantu-Mansy auton. district	532	Ivanovo region	39
Sakhalin region	307	Kirov region	39
Tatarstan Republic	158	Altay region	37
Komi Republic	164	Kurgan region	37
Tomsk region	113	Bryansk region	31
Astrakhan region	109		
Regions of metallurgy		*Republics of South (Northern Caucasus)*	
Vologda region	158	Dagestan	40
Lipersk region	118	Kabardino-Balkaria	39
Agglomerations		Nothern Osetia	35
		Adygeya	30
Leningrad region	197	Ingushetia	15
St. Petersburg	133	*Siberia and Far East regions*	
Moscow	116		
Moscow region	100	Buryatia republic	42
South		Primorsky kray	38
		Tyva Republic	16
Krasnodar region	103		

* with corrections for regional price coefficients.

with the economic situation in the region. Only the inequality of salaries in the regions stopped growing; this is a result of salary increases for the non-market sector (government paid) employees and also due to the increase of social transfers for the people. However, the social aid reform of 2005 resulted in non-targeted social transfers. People who are not poor receive the bulk of the aid, and as a result of this, regional differences in the level of poverty are once again growing.

The dynamics of the inequality indicator show us once again that economic growth can intensify social differences between the

Table 4: The Gini Coefficient for Regional Inequality

	1999	2002	2003	2004	2005	2006
GRP per capita	0.250	0.282	0.283	0.333	0.348	0.333*
Unemployment rate	0.165	0.244	0.257	0.264	0.261	0.281
Incomes	0.253	0.179	0.192	0.194	0.191	0.184
Poverty rate	0.211	0.177	0.176	0.168	0.181	0.181
infant mortality	0.118	0.106	0.105	0.102	0.119	0.132
Life expectancy at birth	0.015	0.018	0.022	0.024	0.026	0.022

* The decrease of the GINI coefficient for GRP was caused the decrease of the number of regions after 6 under developed autonomous districts became part of more developed regions.

regions. This is not just a Russian problem; the same problem persists in all developing countries as well as in some developed European countries.[2] In Russia, a huge oil and gas income is distributed to small groups of rich people from the leading regions, firstly Moscow. This situation adds to the problem.

Big Business Cities

There are three groups of cities in contemporary Russia that are centers of growth. The first and most successful type are the agglomerations of Moscow and St. Petersburg, which are followed by other agglomerations with more than a million people. The second are more or less significant regional centers, which are helped by their status. The third group are monofunctional towns of the export industry (centers of oil and gas production, other export resources and basic refining). These towns develop thanks to companies that belong to Russian big business.

2 P. Martin, "The Geography of Inequalities in Europe", Swedish Economic Policy Review, 12 (2005), pp. 83–108.

There are about 150 such monofunctional towns - one eighth of all cities in the country. Most of them have plants of large private companies; about 15% of them are towns of natural monopolies and large companies under government control (РАО ЕЭС - the power utility monopoly -, Gasprom, and others).

The socio-economical conditions of monofunctional towns differ. Among them there are more prominent ones that hold the main (basic) operations of big businesses; are not that many of them, however - 6% of all the cities in the country with 7% of the population. Another 6% of monotowns stem from the last expansion of big business. Their production is usually smaller and they are undergoing industrial modernization. It is not uncommon for them to lay off employees as they attempt to minimise production costs. There are as many expansion-created monofunctional towns as there are basic ones, but they are one and a half times smaller; only 4% of Russians live in such towns. In this way big business in Russia influences the economic condition of every ninth citizen in the country.

The towns of natural resource monopolies are the smallest (РАО ЕЭС, "Gasprom"). It is a Soviet legacy: around large power plants or gas excavation sites, towns were built to house the workers. During the crisis of the 1990s it was a lot easier and less expensive to provide social services to these small towns. However this advantage has a negative side effect - the entire social policy of large companies becomes corporate since most of the town's residents work for the company.

The differences in concentration of monofunctional towns of big business are very significant. At least half of the regions do not have such towns. Only every fifth Russian region, especially metallurgic and oil producing regions, have a heavy concentrations of these cities. The Sverdlovsk Region has the maximum number of monofunctional towns belonging to big business.

"Basic" towns of big businesses which hold key company resources offer higher salaries (1.5 to 6 times more than the average for the region per month) and they also have more budget income. Thanks to the rise of salaries, consumer behaviour and the way of life is becoming more modern for the citizens. However the development of towns with monoprofile economics can not be stable. Another problem of these towns is the quality of their human resources - the educational level of workers is usually low, with the exception of towns working for the defence industry (closed towns).

Nevertheless, the "basic" towns of big business play a very important role in contemporary Russia. Their development is accelerating the diffusion of modernization from large city-centers towards the periphery. But the direction of this development depends on the politics of big business. If big business and the government were to invest in human capital with the help of various educational and social aid programs, other sectors of the local economies would be able to develop in a stable way. If the only goal is to maximise income, then the town will remain just a factory settlement; this problem has not yet been resolved. Regional differences in politics and business are huge: In the oil extracting autonomous district of Khantu-Mansy, positive results from the cooperation between business and government in social policy can be seen; at the same time the social sphere of many cities in the Urals and Siberia is in a crisis.

2. The Evolution and Territorial Expansion of Russia's Big Business

The evaluation of development in post-Soviet Russia is based on political tendencies and economic dynamics. Another reliant factor is the change in business strategy through time and its territorial configuration. A very good example is the evolution of big business, actually created by the government due to the privatization of the 1990s. At first glance, the history of big business's development in Russia seems like a zero-sum game. The weak state of the 1990s was dominated by business (state capture), but in the year 2000 the state became more powerful and began to suppress (not only Russian) business. However, this banal example simplifies reality. The evolution of business in Russia and the evolution of its relations to the government is better seen through diverse examples of regional practices.

Dividing and Keeping Out

In the beginning of the 1990s it was easy to get into business; the right connections, mobility, and entrepreneurship were often more important than money. In Moscow, with its huge market and concentrated financial resources, businesses emerged and evolved very dynamically. Banking groups were the first to be created. At the same time, in 1992–1993, the process of segmenting government property in the natural resources industry began. Huge oil companies, both regional and national, emerged. These were headed by former Soviet managers but hardly any of them were effective.

By the middle of the 1990s the young banking business began flexing its muscles and, with the help of insider investors, managed to obtain the best government properties in the natural resource industry. In the regions, shares were bought in an organised fashion. The industrial empires of large banks slowly began to emerge. The weak federal government was divided between various warring clans. These were connected with business groups and influenced the process according to their personal interests.

The regional governments were opposed to this property redistribution. Using the weakness of the federal government as an opportunity, the regions obtained maximum political power and also the right to control large assets on their territory. This regionalization process led to the creation of oil companies КомиТЭК-KomiTEK, Татнефть-Tatneft, Башнефть-Bashneft, and many others. By the middle of the 1990s a system of administrative entrepreneurship was formed that was different from the system of administrative rent in the way that bureaucrats gained access (either directly or indirectly) to some of the most prosperous assets. The first collision of big business and government happened in the regions, where the local elites resisted relinquishing their assets, and not in the Kremlin. This conflict continued during the second half of the 1990s. The regional elites were defeated when their election block (All Russian – Motherland) lost the state parliament elections in 1999. This strengthened the federal center and opened the door to the regions for big business.

Rational Alliances and Strategies

Businessmen are able to make rational decisions more quickly than politicians. The financial resources of businesses increased during the period of economic growth and were used to obtain new assets and land. Putin's reforms weakened the positions of regional governments; thus politicians strengthened this process. By 2003 huge Russian corporations were doing more for the country's economic integration than the actions of federal authorities. For example, the largest business groups (the Alpha-group, the Rusal/Bazovij element, and others), bought assets in more than 20 regions; most important to them were the regions where their main production took place.

The various regional strategies of large business groups do not only depend on the peculiarities of a certain business or financial resources but also on the rational choice of the companies' owners. Basic Element Company to this day prefers territorial expansion and maximum diversification; that implies choosing an aggressive development strategy. Alpha Group's regional policy is close to that, but at the same time the group will sell assets if it receives a good price for them; this strategy is known as investment banking. Other groups, including those with more advanced management (Yukos, Norilskij Nikel) and more traditional and specialised oil companies (Lukoil, Surgut Oil and Gas), chose the strategy of concentration in one region and selective expansion, usually into their new resource zones.

The territorial expansion of large Russian business in the 2000s was done along several paths:

- *Economically developed but "closed" regions* (these are completely under the control of the regional leader or his business group) with attractive assets for big business in the oil industry or service industry (Moscow, the republics of Tatarstan, Bashkirtostan, and Komi).
- *Regions with potential resources,* especially oil and gas (Nenets and Evenkisk autonomous districts, republic of Yakutiya) and to a lesser degree coal (regions of South East Siberia and the Far East).

- *Regions with refining industries:* large automotive factories, chemical plants which manufacture goods using resources or materials of Russia's biggest companies.
- *Regions with export sectors that have not yet been divided among the largest business groups:* regions of the lumber and paper manufacturing industries.
- *The most vital transit regions,* through which goods for export pass (Leningrad, Kaliningrad, Krasnodar, and Primoriski regions).

The business groups themselves gradually became large industrial companies. They revealed the names of their holders and improved management in order to be able to enter international lending markets and raise capital.[3]

Pragmatic alliances between regional administrations and corporations were formed. Putin's "vertical power" and limitations on the authority of regions forced the governors to seek new partners that would be able to compensate lost resources. As for big business, it attempted to obtain ownership guarantees in the regions. With this in mind, businesses rationally chose the strategy of working together with the governors. In those regions where agreements were not reached with the administration, they conducted the policy of exporting capital (with the help of exchanging pricing for profit minimization); only the municipality where a basic enterprise was located was supported. A classical example of such relations is Irkutsk region. If arrangements for the protection of business activities and favourable conditions for development were reached, the current governors were supported. If a governor appeared disloyal, large corporations supported his political competitors, using the more costly mechanism of elections. Election campaigns in sparsely populated autonomous regions were the cheapest. A series of victories in governor's elections by candidates supported by proprietors and top-managers of large companies in 1999–2002 drew wide attention, but

3 See J. Pappe. "Change of the Basic Subject in Russian Large Business and Factors Defining It: Results of 2000–2005". Carnegie Moscow Center, 2005 www.carnegie.ru/ru/pubs/media/, and also: J. Pappe and J. Galyhina, "External Economic Factors of Large Business Transformation in Russia", Questions of Economy, 10, 2005.

political forecasts of the mass arrival of businesses to power proved to be inaccurate. Big business, having supported "their" governors in elections, reduced costs - they received tax privileges or blocked the way to prospective resources to competitors. But such a strategy was effective only in small regions; in other cases, the expense of such projects was excessively high. Sometimes such expenses had to be accepted, for example, if there was a risk of "privatisation" of a region by a competitor (as in the struggle between "Norilsk nickel" and "Rusal" in the Krasnojasky region at the beginning of the 2000s). As a result, Russia witnessed the formation of only 5–6 "corporation regions" completely supervised by large business.

The interaction between business and regional authorities strengthened both sides: the regions received additional revenues for their budgets and investments, and businesses received protection of their assets, control over resources and tax laxity. Though such informal arrangements infringed upon the interests of the federal centre and other regions of the country, which as a result received less budgetary resources, it did not exceed the typical conflict between the interests of a part (separate companies and regions) and the whole (country).

The federal authority tolerated the alliances of large business with governors. To put it more precisely, it did not have sufficient resources to interfere. And it was only the case of "Yukos" and the Evenkiysky authorities, who tried to appoint one of the top managers of the company to the upper chamber of the Russian parliament (The Federation Council), that led to the decision of the Kremlin to destroy "region corporations". This was accomplished by uniting some autonomous regions with parent territories. Besides, the possibilities for an alliance between large private businesses and regional authorities decreased after the law limiting the size of regional privileges under the profit tax was passed. Additionally, the abolishment of direct governor elections made them less reliable for and "protective" of their supporters.

It would seem that the "special" relations of business with regional authorities were deprived of an institutional base. But it remained necessary for businesses to protect their assets and influence in the regions; that's why business began to search for other ways. In addition to creating confidential relations (an important resource in Russia), economic levers were widely used in order to influence governors. The most widespread forms of these remain profit understating (a tax on profits is basically allocated for regional budgets) and shifting the legal address of a company to another region. "Special" relations became less important, but the mechanisms of influence remained.

Globalisation and Regionalisation

In 2002–2003 large Russian corporations, having accumulated the necessary financial resources, began global expansion. In the world markets, all needed the political support of the federal authorities. And to some degree, such support was given not only to state companies, though at times the federal authorities behaved aggressively.

The alliances of large private business with regional authorities will inevitably weaken. Today Russian corporations operate not only in national but in global categories, dividing "the centres of profits" and "the centres of expenses". The interests of the regions where basic enterprises are located are less coordinated with the global strategies of large private businesses. Contradictions have to be eliminated for the sake of the stable functioning of enterprises, but in the future the role of out-of-date actives for globalised companies will inevitably decrease. And though it is obvious that large businesses still have to build special relations with regional authorities, in the long run their obligations to regions will decrease. In this author's opinion, relations will smoothly adapt to changes: actors that are comparable in influence will be compelled to "adapt" to each other. The regions have already become a platform for the development of compromises between corporations and governing bodies; such experience is invaluable for a country burdened with authoritative traditions. However, for the state companies of the oil and gas sector, patronised by the federal authority, regional power is not a significant factor: they solve their problems in the Kremlin.

One more tendency is the growth of medium-sized trading businesses in regions. These companies are not uniform. The companies that own old Soviet enterprises are more closely connected with regional power than others. They often prefer to exist in a noncompetitive environment. The market transformation of such companies can occur only with a change of proprietor. Especially in European Russia, the expansion of capital from Moscow serves as a motor in the branches of food and mechanical engineering.

The new businesses that are concentrated in the service sector are more independent, with the exception of the construction businesses of large cities, which are almost everywhere under the control of local authorities and the regional trading companies. The domination of regional business that "accrete" with local authorities is being eliminated by external expansion. The large construction companies are moving from the capital to regions. The big Russian and international trading networks have already taken over almost all large cities of the European part except for the north and the Caucasus republics; expansion has also slowed down in the east of the country because the price of logistics rose. Hotels, dealer networks, and insurance companies (after population incomes grow) are following them. Businesses still enter the region with the help of bribes, but this, rather than being the administrative business of regional authorities, is a corruption rent that kills competition.

The diversification of business in regions promotes not only economic development and the growth of budget income but also reduces the risk to economic (fluctuations of the prices for raw materials) and political conjuncture. Supervising a large number of medium-sized businesses by power methods is more difficult than a small number of large "oligarchic" groups. Growing medium-sized businesses that are implanted in a region are important for another reason as well. In the 2000s, regional policy accelerated the recruitment of local personnel and began to defend local interests. In the election cycle of 2007–2008, it was expected that a mass change of "long-living" governors would occur; regional business representatives of were often considered as alternatives. However, because the governor's elections were cancelled in 2005, the majority of these governors remained in power. Having lost the opportunity to obtain the position of governor, representatives of regional medium-sized businesses became more active in elections of regional legislative bodies and in elections of mayors of large cities. However, the political field has become narrower; a number of regional legislative bodies are elected from party lists and fewer mayors are directly elected.

Having rationally evaluated these current changes, representatives of business have joined parties in elections for regional legislative bodies. The choice of a party banner is quite rational; the sharpest competition is for administrative resources of "Uniform Russia". The second way, executive power in a region, is also rather popular. In the Kaliningrad area Muscovite George Boos, when appointed to the post of governor, at first confronted regional business but quickly reoriented and involved leaders of the regional business community in executive structures. Defending its interests, regional business uses all possible means to exercise influence in the public administration structure; against the background of the absence of capable structures of a civil society, this shift should be considered as a positive one, despite all costs and compromises of such a way.

As for the political role of small businesses, it is clear that small businesses are anxious to survive. Their share has not grown for a long time even in the biggest cities, where 30% of the working population are employed in small businesses. In the south, employment in small businesses reaches 20–25% (taking into account businessmen without registration); however, a considerable part is occupied in "black" economy and consequently very vulnerable to the authorities. If the institutional environment for small businesses is not improved, their role in the economic development of regions and politics will remain marginal.

More State, Less Business

At the beginning of the 2000s it seemed to many that the country was following a normal way of evolutionary development, rewarding everybody: the state provided modernisation of the institutions, and business received stable and predictable rules of the game and constantly grew, paying more and more taxes to the state. But this stability was elusive.

The policy of nationalisation of large businesses began in 2003–2004 with the destruction of the company "Yukos" and the transfer of its major assets (the company "Yuganskneftegaz") to the state-owned company "Rosneft". The property that was bankrupt was transferred to this company only. Later the change in management of the largest automobile company "Autovaz" due to the pressure of public administration institutions resulted in the new owners being the state-owned company "Rosoboronexport". The rest of the companies were bought: some for a very low ("incorporated machine-building factories") others for a very high sum (the purchase of "Sibneft" by Gazprom). The policy of nationalisation was justified by the necessity for concentration of strategic enterprises in the hands of the state, particularly oil and gas corporations, but also metallurgical ones: "VSMPO-AVISMA" was also obtained by "Rosoboronexport" for less than its market value. In June 2007, prime minister Putin approved a list of 480 enterprises which would be transferred to the same state company. In all probability, repartition will continue in the oil sector; experts name the multinational corporation "VR" as the next candidate.

Recently, the state turned to the largest western companies which extract gas on Sakhalin within the framework of the agreement on division of production signed in the 1990s. The attempt to strengthen state control over big business is obvious.

A. Auzan explained that the reason of such renationalisation of property is due to the change of elites.[4] But the "zero-sum" game (the winner takes all) stops the modernization of Russian big business and reduces its efficiency. Without the prospect of stability, businesses will limit their investments into the modernisation of the enterprises. Large corporations will export profits, buying up foreign actives and reducing the risks of confiscation.

Repartition under the slogan of strengthening of the state is approved by the majority of Russians who perceive that the privatisation of 1990 was not legitimate; it is difficult to argue with that. Big business, tied hand and foot by the negative attitude of the society, does not render resistance to property repartition. However, in 1917 it was the same and done without any mortgaging privatisation.

To be engaged in private business is unfashionable today. Employment in state-owned enterprises and in public administration has grown by more than 100% under Putin. The young generation aspires to become officials or to be employed in the large state-

4 A. Auzan, "Ashes are Dispelled Downwind". Novaya gazeta 11, 14 February 2005.

owned companies. How long will all these negative tendencies for business last?

The Forecast for the Future

The big businesses that appeared in Russia in the 1990s developed along a twisting path from a section with the Soviet assets dominating not only economically but also politically, to boundless loyalty to the federal authorities. The problem of the privatisation of the state by "oligarchs" is no longer a pressing one; the country is moving towards state capitalism in all major branches of the economy.

But the "pendulum" is still swinging and the author feels that conditions for a change of direction are gradually being formed, mainly in the regions. Russia cannot be a space in which the general game rules are dictated from the centre – the diversity of the huge territories is too great. The interaction between business and regional authorities shows both negative examples and examples of constructive partnerships. There are objective factors expanding the sphere of normal partner relations.

First and foremost, there are changes in the investment policy of regions. Under the pressure of "vertical power", their resources were reduced in comparison with the federal budget, and "special" relations with big business have become less steady. Regional authorities have to prepare for new investments; this is a direct way to improve the investment climate. Attention to investors is growing, though most regions are searching for large and foreign investors. However, in the Perm region, for the first time all the actors – not just the "especially trusted" and big investors – have received regional privileges under the profit tax. In some regions, authorities, providing concord of transaction, help investors to overcome administrative barriers. And though in many regions a corrupt "entrance fee" is still required, the country is pretty diversified.

Secondly, business in regions has not only grown but also diversified in the last 10–15 years: the number of the middle-sized and local companies has increased. The configuration of businesses of different sizes differs strongly in regions, but diversification has increased everywhere. The ground for informal relations and special preferences to the some companies has been objectively narrowed while the requirement for transparent game rules for businesses is growing. The state formally declares them, but is too keen on sharing assets to watch their implementation. In the regions it is easier to achieve an improvement of the institutional environment if regional authorities have real interest in such an improvement. Such interest is already shown - in particular, in the Tomsk, Perm, and Samara areas - though presently there are few such regions.

Thirdly, the risks of "oligarchical" business do not yet extend to less large companies, especially in the new service sectors of the economy. Certainly, the state (to speak more precisely, concrete officials representing it) can absorb some risks, but not all at once. Given the Russian conditions, there is some guarantee of survival and development. After all, enterprise activity in the country remains high.

But even the growing needs of business for transparent game rules and the occurrence of regions with advanced institutions alone are not guarantees of change. Reforms in Russia have always begun from above.

Dmitrij Furman

2008. New Alternatives in Russian History

In the history of a country, as well as in a biography of an individual, long periods when there are little or no chances to change the key parameters of existence take turns with shorter periods when random combinations of circumstances and a free choice of people can define these parameters for many years ahead. In the history of Russia, the period of "reorganisation" was one of these periods, when much depended on personal and casual factors. The year of 1991 was probably the richest in alternatives: it was the year when Yeltsin was elected president of Russia, the year of the failed putsch of the GKCHP, and the year in which both the allied and Belovezhsky agreements were signed. Certainly, even during such critical epochs, the range of possible parameters for the future life of a country is not boundless. I think that in 1991 Russia had no chance of returning to "Soviet power" and socialism and little chance to choose a way that could have led quickly to the construction of a consolidated democracy with a rotation of different political elites and parties, as happened in the Baltic and Central European countries. But the range of possibilities was all the same enormous - from the preservation of the Union in another form for 10 or more years, to a bloody war between republics as in Yugoslavia, and, for Russia alone, from almost a democracy with a dominating party to an open military dictatorship. After the Belovezhsky agreements, the range of possibilities was sharply narrowed and the next unpredictable situation that was rich in alternatives (though not as rich as in 1991) appeared only in 2007–2008.

* * *

The Belovezhsky agreements are a pivot point in our history, as they brought to life Russia's modern political system of an "imitating democracy". They mark the end the unstable period of dual power - Russian vs. unionist or Yeltsin vs. Gorbachov - and Boris

Yeltsin, who had become president of Russia as a part of the USSR earlier in 1991, remained as the single leader of a democratic anti-Communist movement. The cultural and psychological immaturity of Russian society, which did not have any actual experience of living in democracy (the experience of 1917 was chiefly traumatic and generated fear of democracy), created ample opportunities for the transformation of a proclaimed democracy in a form that covered a real authoritative content, as has happened in ten or even hundreds times in many different countries in modern history. But the way Yeltsin came to power, paved by the Belovezhsky agreements, made any other option of further development not only very improbable, but actually non-existent.

The collapse of the USSR following these agreements was carried out by Yeltsin without a "national mandate". (Only the Ukrainian participant of agreements, L. Kravchuk, had such a mandate, in the form of a decision of a referendum on independence.) Moreover, Yeltsin directly contradicted the results of the referendum about the destiny of the USSR that was held in March of 1991. It meant that now the opposition to Yeltsin and his democratic colleagues had the opportunity to accuse them not simply of the wrong policy, but of illegal actions, destruction of the state and national treachery. And this in turn meant that thereafter the preservation of the presidency by Yeltsin and non-admission of the opposition to power became a categorical imperative for him. He could only leave the post after securely transferring power to a successor appointed by him. It was no longer a question of the desire to implement a certain policy or to enjoy power, but of the preservation of his freedom and possibly even his life.

Yeltsin and his colleagues could not refuse the principles of democracy proclaimed by them, but they also could no longer implement them. The Belovezhsky agreements closed for Yeltsin a possibility which Landsbergis, Valens, and Kravchyk still possessed: leaving power. The construction of an imitating democracy of the "third-world" type – the transformation of power into a "non-alternative" democracy – was also no longer an alternative for Yeltsin and his colleagues. And as the construction of such a system by definition means a regular infringement of the proclaimed democratic principles (that is, the implementation of illegal and anticonstitutional actions), departing from this way became more difficult with each step. If it had still been possible to imagine a situation after the Belovezhsky agreements in which Yeltsin conceded power to an opponent and not to an appointed successor, it became impossible after the parliamentary elections of 1993. Defeat in elections after 1993 would automatically have meant imprisonment for Yeltsin and his colleagues. The winners would take all.

And the features of Russian mass consciousness make this way easy and convenient. Russian society possesses a very weak self-organising ability and a huge "fear of freedom". The imitational democracy that grants personal power to a "non-alternative" president corresponds to the stage of its development. A prospect of free elections which could end with the victory of the opposition was unacceptable not only for Yeltsin but also for the majority of people; this not because they support the policy of their authorities, but rather because of their fear to be left without consolidated power and plunge into anarchy and chaos, and, finally, the fear of returning to "the time of trouble" of 1917.

The political development is thus governed by a rigid logic, derived from the nature of the imitational-democratic system which supports only a small range of variants. Paradoxicallly, any victory achieved in this system narrows the spectrum of choice for the winner and makes his further actions "programmed". And although there was enormious power in Yeltsin's and then in his successor's hands, it was the power over concrete people, a possibility to ennoble and enrich, or to overthrow and ruin them.[1] Henceforth, the country's development depended on them only to a minimal extent.

1 In some aspects this power is absolutely grandiose and even exceeds the power of the emperors and the Communist leaders. After the adoption of the law on succession to the throne, tsars could not define their successors at all and a tsarevitch successor automatically became tsar. The Soviet Secretary-Generals also could not define their successors. These were chosen by the Politburo, and appointments for higher posts were limited by a narrow circle of Politburo. Appointments of well-known persons to high positions such as the head of government were impossible in the Soviet system. On the other hand, Yeltsin and Putin could appoint practically everybody as successors. Putin's appointment was absolutely unexpected while Putin's officials – Fradkov and Zybkov – were unknown officials and even in Putin's inner circle, there was nobody, I am sure, who could foresee that the president's choice would fall on them.

A detailed description of all aspects of this logic would exceed the scope of this essay; I thus will limit myself to a short explanation.

Expansion of the non-alternative sphere in politics.[2] The necessity for supporting non-alternative presidential power assumes the consecutive expansion of a sphere of control over political life. At first, the president simply confirms his power and eliminates real opponents and direct threats to his power. But then he has to watch that no new threats arise and stop those as soon as possible.

First of all, the establishment of the system of "non-alternative" presidency means a conflict with other branches of power and their submission, that is, an actual liquidation of the division of power. This happened in 1993 when Yeltsin dispersed the parliament with force (with very weak public resistance) and then organised the adoption of the constitution which suited him best. Any constitution is inconvenient for the personal power of the president, but it is impossible to live without any constitution.

This adoption of a convenient constitution that gives the president enormous power and sharply limits the rights of the parliament is an important first step. But naturally others steps have to follow. Even restricted by the new constitution, a weak parliament that is led by the opposition is dangerous. Therefore, it is necessary for the president to establish control over all the selection processes so that he will securely obtain the "correct" result - a parliament that is not hostile to the president and supervised by him. And this assumes in turn:

1) controlling regional and local authorities which will provide the necessary election results,
2) controlling the party system; that is, only a fictitious or weak opposition,
3) creation of an own party and support of its leading members,
4) controlling the mass media, first of all TV,

2 See in more detail: Dmitrij Furman, "Ursprünge und Elemente imitierter Demokratien. Zur politischen Entwicklung im postsowietischen Raum", Osteuropa 56, 9, September 2006, pp. 2–25.

5) establishing the dependence of courts on "executive" power, guaranteeing adoption of "correct" decisions in affairs having a political value,
6) using privatisation for the creation of a stratum dependent on presidential power and proprietors interested in its preservation, and
7) the suppression of attempts of large proprietors ("oligarchs") to play an independent political role.

The primary goals arising during the evolution and strengthening of a system of "non-alternative power", were already solved by Yeltsin, but some decisions were passed on to his successor Putin. However, these problems logically follow one another; had Yeltsin not been a sick man and had he not felt that he had only the choice of fast death in his position or the transfer of power to a successor, he would have continued to reign and himself solved those problems that were later solved by Putin. And if he had not selected Putin as his successor but another person (Yeltsin "rushed about", searching for a suitable successor, and Putin's choice happened at the last minute and was to an enormous degree casual), they would have been solved by this other person. The law of our political evolution is visible in its comparison with evolution of other post-Soviet countries, for example, Belarus or Kazakhstan. Everywhere the same problems were solved, following from the essence of an imitational-democratic system, in a similar image and in a similar logical sequence. And if we take into account that modifications in the post-Soviet development of different countries can be explained to an enormous degree by various objective factors – for example, by features of national cultures, by the different resources of countries, and by their different geopolitical position – we will understand that the subjective and individual role of post-Soviet leaders is not great. Some personal features of Yeltsin and Putin define individual "styles" of problem-solving, but there was always only one possible result. For example, Yeltsin's impulsiveness and roughness played a big role in the bloody form of suppression and dispersal of parliament in 1993. In this case, the form was not inevitable, and other presidents of the CIS dispersing parliaments managed this without blood. But it was impossible for Yeltsin not to disperse parliament, which possessed enormous constitutional powers and considered itself as a principal organ of power and an embodiment of national will. Similarly, at a later stage, the president (now Putin) could establish control of power over TV somewhat differently, not by expelling too independent journalists, but on the contrary, by "taming" them. It would have been impossible for Putin, however, not to establish control over TV.

The idea of opposition between Yeltsin and Putin, which is popular among Russian liberals, is based on a misunderstanding of the difference between the Yeltsin and Putin epochs. They are chiefly differences between the natural stages of development of one system, and only to a small degree connected to the distinctions between Yeltsin and his successor.[3] (The differences between the Lenin and Stalin periods were similar; they were differences of stages of development of the Soviet system.)

These distinctions of individuals are connected with the laws of development of a system. In many aspects Putin is a psychological contrast to the impulsive, unpredictable and often drunk Yeltsin. In order to solve the new problems of the following stage, a stage of consolidation and stabilisation of the regime, Yeltsin chose a person with other qualities than his as his successor, just as at another time party leaders chose Stalin, a person with absolutely other qualities than Lenin, as his successor in order to solve the "new problems facing the party".

Socio-economic development and its rhythm. In my opinion, in post-Soviet social and economic development, the natural prevails considerably over the casual and the personal. A specific form of privatisation is natural – under "non-alternative" power it cannot be

3 This misunderstanding also has a certain psychological basis. Modern Russian liberals are basically people who held important posts at the beginning of the 1990s and then were discharged of power or left on their own when they saw that the system was clearly becoming more and more authoritative. For them to recognise that the present regime is the result of a natural evolution of the system that appeared in 1991 is equivalent to the recognition that they participated in its creation and bear responsibility for it. One can compare them with those old Bolsheviks who carried out the October revolution and terror against ideological opponents of Bolshevism, but when Stalin aimed the same terror at his inner-party opposition, they were terrified and began to blame him for having departed from the "norms of party life" of Lenin.

anything else than the actual distribution of large state ownership, enriching the reigning group and creating a layer of proprietors directly interested in power and depending upon it. It is natural to use quasi-legal methods of control over proprietors to encourage political loyalty and to ruin disloyal elements. The establishment of direct control by power over the most profitable and economic key industries is also natural.

The natural is a rhythm and a cycle of our economic development. The transition from socialism to an economic system based on private property and market economics could not but be accompanied by a collapse of manufacture and a sharp reduction of the standard of living for the majority of the population. This happened in all post-communist countries. In all of these countries, the fall was also followed by a rise. The most basic sources of this rise were the adaptation of individuals and society in general to new forms of economic life, the formation of new skills and habits, and new forms of economical organisation. In Russia the rise, which followed a recession, was facilitated by an external and - in relation to Russian post-Soviet development - casual circumstance: rich stocks of oil and gas existed and world prices for energy carriers increased sharply. But the general rhythm of rise and fall does not depend on these external circumstances, as we see the same rhythm in post-communist countries not only in oil and gas exporting countries such as Azerbaijan and Kazakhstan (where the rates of GNP growth are even higher than those of Russia), but also in countries which do not have oil and gas at all.

Therefore the perception of a contradiction between the roles of the "democrat" Yeltsin and the authoritative ex-KGB officer Putin is the same as assigning responsibility for the devastation at the beginning of the 1990s to Yeltsin and for the following upsurge to Putin; their differences are the differences of stages of development within the same system. It is natural that the upsurge *under* Putin is perceived as the upsurge *because of* Putin in the national consciousness, but it is the same logic which ancient and primitive people used when they attributed good harvests to the virtues (or even their sexual prowess), and invasion of locusts to the sins (or the weakening of the sexual prowess) of their local rulers. If Yeltsin had lived and had been well, or if he had chosen not Putin as a successor but Aksyonenko (already forgotten by now, he was the minister of transport whom Yeltsin almost made to his successor, changing his mind at the last moment) or someone else, the upsurge would have been under them. And now people would be surprised at Yeltsin's wisdom, which led the country through all difficulties to an upsurge, or would oppose Yeltsin's vicious course, under which people could hardly make ends meet, and praise the wise policy of Aksyonenko, under which they, at last, felt relieved.

The Evolution of Foreign Policy: From "Lowering to the Knees" to "Lifting from the Knees"

One more myth in the Russian (and not only the Russian) consciousnesses that is connected with the "personification" of natural stages of development, is the myth that one merit (for others it is not a merit but rather a fault) of Putin is that it was he who conceived the new confrontational line towards the West. (In the Russian patriotic lexicon it sounds as follows: Putin, having ceased "to fawn upon the West", "has lifted Russia from her knees" and began "to defend Russian national interests".)

There are many fewer differences in the relations between Russia and the West under Yeltsin and Putin than is now frequently represented. Intervention in the affairs of neighbours - ex-"brother republics" - was more direct and rougher under Yeltsin than under Putin, though the West treated these interventions absolutely differently. Yeltsin's reaction to the Yugoslavian events was a threat to use nuclear weapons and a march of the Russian armies to Prishtina, and counteraction to NATO expansion was at not all less sharp than now (if Putin now states that Ukraine's membership to NATO could lead to the aiming of Russian rockets at Ukraine, Yeltsin's Minister for Foreign Affairs Primakov declared that if the Baltic States entered NATO, Russian external and military policy would be revised entirely). The contrast between Yeltsin and Putin is mainly explained by a distinction of perception - the West perceived even sharp and aggressive actions of Yeltsin as being casual or compelled by the necessity to consider attacks by the communist and nationalist opposition, or that they originated not from Yeltsin but from some "reactionaries" in his surroundings. Nevertheless, distinctions in the foreign policy of Russia under Yeltsin and Putin undoubtedly exist. Under Putin, Russia more actively and more often resisted the West on different questions and in different regions, but basically only on post-Soviet territory, where there is even something like a local "cold war" of weak intensity. But these changes are only vaguely connected with the individuals Putin and Yeltsin.

In 1991, the declaration of a new independent, democratic Russia with a market economy was accompanied by euphoria in Russian–Western relations. This statement does not need to be substantiated. It is also clear that subsequent Russian evolution could only be accompanied by a deterioration of these relations and the re-occurrence of elements of the Cold War. During this evolution our system became less similar to the western system. If at first distinctions between Russian and Western societies could be perceived by Western (and Russian) politicians and analysts as a display of the underdevelopment of Russian society which, however, was following a correct way and would gradually approach the Western democratic model, it later became obvious that some problems were not due to this underdevelopment, but that the political system had developed in another direction. This direction of development, which for Russia excludes the possibility of integration into western structures, caused dissatisfaction in the West (especially because of the euphoria that had preceded it) and in the Russian government, which naturally is not satisfied with low rankings in Western ratings, and Western criticism cannot change the direction of natural development of the Russian political system.

In such a situation the growth of conflicts in the relations between Russia and the West and an amplifying aspiration of Russia "to rise from her knees" and to resist the West are quite natural. If we understand that Russia - having inherited from the USSR the status of being a superpower that plays an active geopolitical role, possesses nuclear weapons, and has permanent membership in the UNSC - has been weakened - having had to overcome a crisis connected with a "formational" transition, having been exposed to an increase in world prices for oil and gas, and having become much more financially dependent upon the West –, we will understand that Putin's individual role in the transformation of Russian foreign policy is not great. As the building of "imitating democracy" after 1991 went naturally on its internal logic, just as a post-transformational upswing naturally follows a transformational economic recession, it is also natural that a period of mutual disappointment and a soft and non-ideological revival of the Cold War should follow the euphoric period of the Russian–Western relations. Foreign policy changes are a projection of the internal evolution of a society and are subordinated to the uniform logic of the Russian post-Soviet development.

We can trace similar natural evolutionary processes to the most diverse spheres of public life. For example, it is natural that an authoritative political evolution is accompanied by a change of ideological accents; there cannot be full refusal of the democratic principles proclaimed in 1991, for it would mean delegitimation of power. But the symbolical continuity with the imperial and Soviet periods[4] and the accentuation of the Russian cultural originality is strengthened. A sharp distinction between the Russian and Western political systems is comprehended as a "civilised" distinction, western criticism and pressure are seen as attempts to destroy our "civilised kernel" and impose other cultural models. In a religious sphere, the transition from the broadest religious openness at the end of the 1980s and the beginning of the 1990s, to the statement about an official role of Orthodoxy – a symbol and keeper of the Russian "originality" –, and the struggle against the spreading of new religious currents in Russia that destroy this originality, corresponds to it. These natural processes even cover such spheres as sexual morality and fine arts.

Even the rhythm of popularity of the Russian governors (Putin, undoubtedly, is much more popular than Yeltsin was) depends more on the natural rhythms of development of our system than on their personal qualities. An economic upswing, the stabilisation of the imitational-democratic system and a partial return to the habitual

4 The symbolical continuity with an imperial period was actually affirmed from the very beginning of post-Soviet development with the acceptance of a two-headed eagle as state emblem, naming parliament the "Duma" etc. Later, especially under Putin, the continuity with the Soviet period was underlined: this continuity is only partial, selecting the idea of being a "great power" – but not the revolutionary and socialist aspects – from the Soviet past.

forms of a political life such as the change of the style of foreign policy, as I tried to show; all these are natural phenomena and psychologically convenient for the majority of people. And it is also natural that the president under whom "life adjusts" and comes back to habitual forms is more popular than the president under whose governance a terrible collapse of the economy and a sharp weakening of Russia occurred. In this situation, any governor whom Yeltsin would have selected to be his successor would inevitably be more popular than Yeltsin.[5]

A description and analysis of this general logic of evolution in all spheres of public life is a challenge. But from the aforesaid it seems clear to me that in 1991 there was something like a "germ" of present Russia, possessing a certain "genotype", in which all its basic lines are recorded. A germ naturally turns to a mature organism which eventually also naturally achieves old age and dies. From

5 It is necessary to note not only Putin's larger popularity, but also a natural change in their social support. Yeltsin's electoral support had the accurate socially-demographic person. Yeltsin and the revolution of 1991 in general were more supported by the inhabitants of capitals and large cities than the inhabitants of provinces; they were more likely to be educated, well-to-do, and young, than uneducated, poor, and old. In Putin's reign interviews and data of electoral statistics discovered a phenomenon that is at first sight surprising – the huge support of the president did not correlate with any socio-demographic characteristics at all (D. Furman, K. Kaaariajnen See, "I Trust the President", Free Thought, XXI, 2003(5)). The role and image of the second post-revolutionary governor in a non-democratic system is ambiguous. He is a continuer of the first governor who led the revolution. But on the other hand, his functions are not revolutionary; rather, his job is to stabilise the system and calm society, and he restores much of the pre-revolutionary authoritative order. These are the reasons for the unusual social support he receives. As a continuer of Yeltsin and the "revolution", fixing its achievements, first of all in a social and economic sphere, Putin preserved support of those social forces which were the centre of Yeltsin's social base. But as the governor under whom the country began to come back to political order, distinctly reminding of the pre-revolutionary order and returning to pre-revolutionary symbols, he obtained the support of a considerable part of those who had been in opposition to Yeltsin. It is very similar to the support the second governor in the Soviet cycle, Stalin, received. As "true ленинец" Stalin preserved support of those social forces which supported October revolution. But as "a repairer of order" and imperial power of Russia he received support from a part of the former "counter-revolutionaries".

1991 to 2008 our evolution was subordinated to an uniform logic of disclosing this genotype which does not differ essentially from the genotype of other post-Soviet systems and a number of "third world" systems; distinctions between these are distinctions of individuals that belong to the same kind. The general direction of evolution does not depend on the features of our presidents and their free or determined evolution, but depends on purely individual personal features of decisions.

But in 2007–2008 a personal factor again played an enormous role. The general logic of post-Soviet Russian development was broken by Putin's decision to resign upon the expiration of his second presidential term as the constitution demands, and to become the prime minister under his successor Medvedev, whom he appointed.

* * *

If in systems of "real" democracy stability is a stability of institutions and the constitution instead of the power of certain persons or parties, then in "imitating" democracies stability is defined by the length of the stay in power of certain persons (or quasi-dynasties of persons, each of whom defines his successor), and the "front" constitution can change if it becomes inconvenient or impedes the preservation of power. Here, there are no constant game rules under which there can be different winners, but rather the constant winner who can change the rules of the game. Such post-Soviet presidents as Nazarbaev, Karimov, Rakhmonov, Lukashenko, and Akaev have repeatedly changed constitutions or modified old ones, adopting amendments to them, always in the same direction – a strengthening of presidential power, removal of restrictions on the stay of one person in power etc. Thus these changes in the "front" constitutions never destabilised the systems of personal power in these countries; on the contrary, they strengthened them. Destabilisation in such systems does not occur due to a change of "form" but rather due to a change of "content" – a change or prospect of change of the head of state. An illness of the president is a considerably more destabilising event than a change of constitution.

Yeltsin reigned under two constitutions. He himself created the second one, as convenient as possible for him, in order to radically change the management organisation while formally not contradicting it. Nevertheless, in this constitution there was a clear point on restriction of stay at power of one person to two terms – when it was accepted, the end of these two terms seemed far. Ill Yeltsin did not begin to achieve a change of this point and retired before reaching the end of the second term. Many other problems connected with the construction and strengthening of a system of authoritative power were camouflaged in a constitutional and democratic form. This problem was left by him to a successor.

The change of the constitution and the removal of restriction of president power by two terms, as it happened in other similarly developing post-Soviet republics, would be a logical and natural step in evolution of our political system. It would correspond not only to natural aspiration for any politician, especially for one as young and vigorous as Putin, to preserve power, but would also correspond to the need of stabilisation of the developed system. It would correspond to the Russian political traditions of undivided power that is limited in time only by death.

For Putin to organise a referendum on a change of constitution would be a very easy business. His degree of control over the society and his popularity is considerably higher than Yeltsin's were, and the necessary results would be achieved even without falsifications. Certainly, it would cause some new frictions in relations with the West, but these frictions would not have any serious consequences (as similar changes of constitutions in other CIS countries did not lead to any serious consequences in relations with the West; Kazakhstan, which has recently implemented such changes, is now going to become OSCE chairman). Hence, Putin's decision to adhere to constitutional rule and to relinquish his position is absolutely unusual (all other country leaders of the CIS chose a different way in similar situations) and undoubtedly contradicts the desires of the bureaucracy and people. It is the first case, not only in post-Soviet, but in all Russian history, of an individual voluntarily leaving the post of governor while not leaving politics – Yeltsin's retirement was obviously connected with his health and was not perceived as a precedent.

The question about his motives for this act has not been resolved. Nevertheless, it is possible to make some assumptions. Putin casually got power, and he doesn't look like a person who dreamt of such power and aspired to it. He perceived his unexpected "good luck" as a "challenge", as work which he should do well. His aspiration to power centralisation and opposition suppression is obviously motivated by a representation about stability and order which the country needs and not by his love of power (in the 20th century Russia exhausted its limit of revolutions). He established stability and order as they could be established within the limits of an already developed imitational-democratic system where the order and stability are not separable from the control of the supreme power over a society. When he says that he "ploughed" for eight years and that he is tired, it is believable.

It is obvious that the constitution is somehow already included in his representation of order. In any case, the aspiration not to break the constitution serves as a quite convenient pretext to be released from work.

In the difficult complex of Putin's motives, some motives connected with cultural features of Russia can be presented. In its ability (or inability) to live in conditions of democracy at the present stage of development, Russia is at the level of many countries of "the third world". At the same time Russia is a country that actively participated in the European cultural process of the 19th and 20th centuries and has belonged to the "concert" of European powers since the 18th century. For Russia it is very important to represent its development as a European variant and it is important for Russia to see that her political system is a variant of democratic western systems (immeasurably more than, for example, China or even Kazakhstan). A change of constitution would mean that there are no serious distinctions between Russia and countries of "the third world" and between Putin and presidents such as Mubarak or Nazarbaev. Obviously, for Putin such a recognition would be too difficult.

But the motives for the actions of Putin (about which we can only guess) are less important than their consequences.

First of all, Putin's departure is a blow on the deep "archetypes" of Russian consciousness generated by centuries of autocracy in frameworks in which there are no living "former tsars", and it is a significant step on the road of "modernisation" of the Russian consciousness and the "desacrilisation" and "depersonalisation" of the supreme power. It is an important precedent of submission of the governor to the constitution, to "a paper scrap", as Russian publicists called it when they attempted to convince Putin to remain, lifting a real value of the constitution and law in general, and making the extension of the reign beyond the limit of two terms practically impossible for succeeding presidents. Presidential power that is limited in time is essentially a new phenomenon in Russian history. But the restriction in time is not separable from other restrictions. The whole system of relations around a governor whose term of power is defined and limited will have distinctions from the system of relations between citizens and the governor about whom it is known that he will reign until his death. Presidential power will also be limited by the presence of living, healthy and active predecessors, who will themselves carry a part of that "sacred trembling" which they provoked as masters of Russia. Russia will not pass from an imitating democracy to a real one by Putin's historical decision, but it now is less an imitating democracy than Kazakhstan or Belarus.

The system becomes less certain, steady, there is an incompleteness of power in it and even there is something like a real division of power. This instability is only aggravated by the fact that Putin chose Medvedev as a successor (for the performance of lawyer Medvedev, invectives against Russian "legal nihilism" and an emphasis on the value of law and order are typical) and by his decision to become prime minister under his successor – to help him at the beginning; but actually this is a decision that weakens "one-man management" and creates something like a "diarchy".

The obvious instability of the present situation means that for the first time in Russia since 1991 there is a possibility of choice between different variants of development. What options exist?

* * *

Regimes of "imitating democracies" are internally inconsistent (the form in them considerably and essentially contradicts the content) and consequently are not too steady and durable. The more formal and predictable elections become (when the legalisature of such a regime is finally democratic) and the more the control of power over society becomes rigid, the more the "feedback" between power and society weakens. The end of such a regime earlier or later, in one form or other, and with those or other consequences, is inevitable. In this respect there are no alternatives, and I think that very few people can believe that in Russia a chain of presidents transferring power to each other will last the entire 21st century. But if there is no alternative to the end of similar regimes, very important alternatives exist that are connected with the form, terms, and consequences of this end.

As it is visible from post-Soviet experience, more "liberal" imitating regimes, whose control over society is rather weak, are less durable than more rigid ones which completely suppress legal opposition. The regime of Kuchminsky was weaker than that of Lukashenko and fell while the Lukashenko regime prospers, the Akaev regime fell while Karimov lives on and is well. And from this it is easy to draw a conclusion – if you want stability of the regime, be more rigid. But this conclusion which was drawn from "colour revolutions" by the majority of post-Soviet presidents, is only partly true. Similar stabilisation is interfaced with big risks as well.

The "colour revolutions" had the character of peace. The opposition under "soft" regimes was a legally operating organised force, capable of attracting broad masses and entering negotiations. Both the government and the opposition recognised the constitution, and the results of elections are easy to control and even to correct. The element of spontaneous violation of laws is minimal in such revolutions. In the regimes of Uzbek and Kazakhstan such variants of development do not exist. The legal opposition is practically not present, elections are completely ritualised and no hopes are connected with them. This prolongs the life of a regime but makes its fall more catastrophic. In the CIS we have seen only one prototype of revolutions in the countries with rigid regimes – the Andizhan events in Uzbekistan, unexpected spontaneous explosion of broad masses, self-organising and acting under radical slogans. But out of the CIS we have seen a number of examples of the "rigid falling" of such "rigid" regimes (a distinction in forms of falling of more rigid and softer regimes can be obtained perfectly by comparing the examples of the overthrow of the Somoza regime in Nicaragua and the falling of the regime of the institutional revolutionary party in Mexico).

A more important alternative is connected not with the form and terms of the end of a regime, but with its consequences. There are two variants of such consequences logically possible; either a country passes on to real democracy, or, after a period of anarchy, a new variant of the same imitating regime is established (as it was, for example, in Indonesia when the Sukarno regime fell and the Syharto regime was established, or as it is obviously now occurring in Kirghizia). Upon what do these different consequences of the end of a regime depend?

The possibilities of transition to real democracy is linked to the mastering of democratic values by a society. The movement to democracy is an aspect of public development and "modernisation", and it is clear that despite the totalitarian character of the communistic regime, the people of the Soviet republics in 1991 were culturally, socially, and psychologically closer to democracy than in 1917. The six-year period of the Gorbachev reforms, preparing people for new forms of life, played an enormous role in that the fall of the Soviet Union did not result in a bloody war of all against all and generate a new dictatorship. The new forms are not forms of a democratic legal society. But in order to estimate their value, it is necessary to compare present Russia not only with the USA or Germany, but first of all with the Russia of the past. For Russia, modern imitating democracy is enormous progress, as it is in countries such as Kazakhstan or Uzbekistan.

If a "psychological preparation" for democracy began under the period of the Soviet power, then mastering of democratic norms took place under post-Soviet imitating regimes. Any imitating regime appeals to democratic values and to some extent promotes fastening of these values in this way. There is no doubt, for example, that modern Russia with its experience of market economy, ideological pluralism (albeit a controllable and limited, but nevertheless real political struggle), is more ready for democracy than the Russia of 1991 was, following an experience of only Soviet power and autocracy. But the more liberal the imitating regime is, the more it appeals to legal rights and leaves a space for freedom, the more it promotes acquisition of democratic skills and habits. Its end will thus mean not an immersal in anarchy from which the country can only emerge by new authoritarism, but transition to real and stable democracy.

This should be applied to the analysis of the alternatives that Putin faced. If everything continues to develop as it has developed until 2008, an inevitable end of the present system in the future, the rupture of a chain of presidents transferring power to each other, would have a catastrophic character. Putin's decision to abide by the constitution does not mean, certainly, that our next attempt of transition to democracy can go in general without crisis. The first time a politician comes to power who has not been appointed will be a crisis. But Putin's decision, especially, if it is supported by the already designated "legal" orientation of his successor, creates the possibility of minimisation of this inevitable crisis; maybe it will be the last crisis before the definitive implementation of democracy.

An unstable situation created by the decision of Putin can have, of course, other outcomes. There can be a further fastening of the legal aspects of the system, its further removal from the "main" direction of development of imitating systems, but there can be a "personal reaction". An opened possibility of a softer development is only a possibility. Whether it is realised will depend on the further actions of Medvedev and Putin and on a lot of other factors, including world oil prices, the effect of which is impossible to predict.

Maria Gordeyeva

The Choice of Russia: The Lure and the Misery of Political Stability

The Russian longing for stability which can be verified in practically every national survey undertaken in the past 10–15 years[1] is by no means something particularly Russian. Stability as a paradigm of social development is becoming more and more attractive worldwide with the advance of such concepts as the "consumer society" and the "social state". Moreover, stability seems to be the right path to overcome or rather sit out many of the current economic and social challenges. On the other hand, a number of dangers arise at the very moment when the "stable" development of a particular society becomes its ideology and philosophy.

Political stability as a strategic tool is a part of every political system because, regardless of the features and the leading ideology, time and the continuity are the factors that give it credibility and thus legitimate the regime. It can be best implemented through a number of democratic mechanisms allowing for the necessary representation and decreasing tension attributed to the inability to satisfy the needs of all social groups. These mechanisms of political representation are probably the most efficient ones when it comes to bringing together various points of view and helping to transfer the existing disputes and disagreements into a civilized public discussion. They also provide forums and public arenas where different opinions can be presented and defended. On the other hand, one should be extremely cautious and take into account the innate conflict between ideal democracy with equality of opportunity and reality, in which this balance is not achieved.

Stability seems to be an essential component of the success of a regime and its proper functioning, but, on the other hand, the discrepancy between the ideal mode of democratic organization and

1 VCIOM, national samples, 1998–2008.

the realistic one is potentially extremely dangerous. It often results in the creation of an imitation of these mechanisms and discussions, thus leading to stagnation and the inability of the system to address an ever-growing number of challenges.

Any social interaction is characterized by the emergence of a number of interpretations of reality; they are perceived as having the power to influence political life around us and change it. Therefore, stability could be best seen as a frame that allows the continuity of such interaction and a variety of paradigms competing with each other. The major problem, which arises almost immediately when authorities turn to "stability" as an underlining concept for their program, is the simplification or mere imitation of the dialogue which should actually exist to counterbalance various interests.

In Russia, "stability" is one of the basic values of society but the term has very peculiar meaning. Russia was trapped in this thinking eight years ago when the country was longing for a break in the constant flow of social and economic experiments leading to turmoil and chaos. The last years were marked by the ambition of both society and the elites to stabilize political power in the country and the existing democracy as such. And, to some extent, stability has already been named the final point of destination in our political and social development. It may seem surprising to us today, but at the turn of the century Russians were ready to sacrifice almost everything in order to feel secure and protected by the strong federal power and the state. The evolution of this notion has brought us to almost the same picture – stability in exchange for personal freedom.

Understanding the unequivocal choice of Russia for stability implies analyzing the phenomenon in the context of three major dimensions: political leadership, political psychology and the dynamics of political development.

The Russian Political System and "Stability"

The stability of the regime and the level of support for a political system are closely interrelated. However, the mechanisms of interaction between these two factors differ drastically in old (consolidated) and new (emerging or consolidating) democracies. In consolidated democratic systems the level of support is high enough to allow the overcoming of various crises and to balance its weaknesses. In emerging democracies, the exaggerated longing for stability and excessive trust in political institutions are more harmful than healing.

Russia presents a clear case of an emerging democracy and fits into all the patterns of political development of the latter. Actually, all of the countries that have emerged on the territory of the former Soviet Union have a strong inclination to exaggerate trust in their political institutions and lack healthy skepticism of their governments and elites.[2] Direct consequences of these realities of political development are public apathy and lack of political participation. Thus, it sounds strange that some analysts criticizing civil society in Russia claim that it is "underdeveloped"; it would be more precise to say that Russia's civil society is "non-existent". The reason for this claim is that there is no distance between society and the power: people associate themselves with most popular leaders, pass the whole burden of responsibility on them and wish to be a part of the system of "vertical mobility" that enables them to move up, all within the boundaries of the governmental system.

Consolidated democracies are, first and foremost, characterized by an effective system of decision-making based on democratic mechanisms, continuity and change in these procedures, and free elections. It is obvious that Russia hardly qualifies as a consolidated democracy; most of these fundamental procedures are only imitated, or they existing without adding any real value to the system. In order to become a consolidated democracy, the following criteria still have to be met:

2 W. Mishler and R. Rose, "Trust, Distrust and Skepticism: Popular Evaluations of Civil and Political Institutions in Post-Communism Societies", Journal of Politics, 59(2), 1997, pp. 418–451.

First, it is necessary that the official mode of power be identical with the political practices and culture of the society. The tension in the system otherwise will make it impossible to coordinate interests of the various groups taking part in the dialogue. On the other hand, the system should not be completely synchronized because it will then result in a totalitarian society. Thinking of Russia in that respect, it is obvious that policies and practices are neither identical nor really support and complement each other. The country seems to operate to a great extent through a number of modified old mechanisms; new ones can be seen only in a number of spheres.

Second, political power should use a variety of procedures and strictly keep the balance between democratic procedures and all the rest.[3] Applying this principle to Russia poses certain difficulties as well; it is hard to determine whether the country has even the required minimum of working democratic credentials.

Russia is still seeking its own formula of stability and sustainable development, which now exists chiefly through the personified trust of particular politicians and leaders (this has been continuously reflected in the national polls starting with the emergence of Putin). This trait is one of the primary features of consolidating democracies; these are usually characterized by passing dissatisfaction with the system on to the leader and in this way triggering systemic changes when often only a minor adjustment is required.[4]

Additionally, Russia's desperate search for stability is predetermined by the unfortunate political experience of the past decades that influenced people either to embrace a new set of individualistic values or to press for the restoration of the old values. Those who attempt to follow a completely different path still experience similar difficulties as in the authoritarian society. And, therefore, in Russia the level of tolerance, the factor that is the basis of many problems of social development, is remarkably low. The most important consequence of this is the considerable lowering of the tolerance of any

3 Harry Eckstein, Regarding Politics: Essays on Political Theory, Stability, and Change. Berkeley: University of California Press, 1992, p. 182.

4 M. A. Seligson and J. F. Carrion, Political Support, Political Scepticism and Political Stability in New Democracies: An Empirical Examination of Mass Support for Coups d'Etat in Peru, p. 73.

social and economic inconveniences. As a result, the elite has to prove its efficiency and has to do it constantly.

Any stable regime can be named as such only if many of the procedures as well as its institutional design can be characterized as long-term ones. The key to long-term existence lies in constant political participation of citizens providing feedback and confirming the maintenance of a proper balance of power in the society.[5] The second reason why this political phenomenon is of utmost importance is because people have to share responsibility for what is happening in the country with the elite, thus helping to reach an acceptable level of coordination of interests and conflict management. This kind of participation can in the first place be implemented through various public organizations and parties. However, Russians can hardly think of using such mechanisms because of a traditional aversion to organizations and parties and a great deal of disrespect for their activists. What is more, Russians these days are most interested in their prospects for improving their own welfare and using stability to ensure the long-term dimension of this process.

It seems that Russia fits some criteria of a consolidated democracy but in some other instances is extremely primitive and backwards. The situation is worsened by the fact that Russia has not overcome the institutional crisis which started with the demise of the Soviet Union but then was transformed into a hybrid existence of several layers of institutions (some are from Yeltsin's time and new ones have been created under Putin). The "crossed wires" of the guiding principles of each of these kinds of institutions, established to serve the conflicting purposes of liberal democracy and control, ruled out any possibility for transparency. Thus, the whole system of public administration has been doomed to confusion, inefficiency and inflexibility. The backbone of this system is definitely a person who could amazingly well connect various parts of it and back the system up by a loyal rotation of officials within its boundaries.

It seems that Putin was to a great extent lucky. It's a sheer coincidence that his measures to pull the country's administration together

5 R. Inglehart, Modernization and Post Modernization: Cultural, Economic and Political Change in 43 Nations. Princeton, NY: Princeton University Press, 1997, p. 148.

were naturally accompanied by the slow emergence of a new social stratification of Russia. To put it mildly, the essence of the current communication between the society and the ruling elite is simple – support in return for stability. The majority of Russians that support Putin believe that he is able to establish personal control over the situation in the country. The trust, thus, is highly personified and results in a significant decrease of public interest for politics and public organizations. That is extremely significant because this disinterest is most widely spread among the people of the middle class, or, using the local modification of the term, the middle layer of society (the term "middle layer" is widely used by the VCIOM analysts to refer to the Russian middle class because they can hardly trace the minimum required features of a classical "middle class" in the country). Members of the middle layer are pushing the country towards "stability by all means" and want the power to make this the most influential "national project". They have purely economic motives – they are interested in stable savings and social benefits increasing annually and "vertical mobility", giving in return their unequivocal support to the ruling elite.

Summing up, it should be pointed out that, even though we can trace a number of features belonging to a society with a high economic and social profile, these are not crucial because the principal characteristics of a balanced long-standing democracy are lacking: stability is personified and personal ratings are immediately transferred to judge efficiency of the system on the whole.

The abovementioned problem is accompanied by a low level of political participation of Russians and their particular distrust of public and non-governmental organizations. Political participation is of crucial importance to any stable regime because it is a trigger and indeed the essence of dialogue between two actors of political communication (meaning the power and the society) and the feedback providing dynamic balance of any long-lasting stability.

Thus, it can be inferred from the analysis of the current situation that even though Russia has a national "demand" for stability, it remains largely unfulfilled at systemic level.

The Perceptional Lure of Political Stability in Russia

Stability in psychological terms means minimizing uncertainty and suspense. This frame is the one most capable of fulfilling our craving for structure and the best one to use when simplifying the various political and social phenomena that exist around us.

Unfortunately, choosing and following the path of stability in its long-term perspective will require a great degree of consolidation of public opinion and implanting the very notion of stability in the set of basic values of any society. Applying this concept to Russian society, it should be pointed out that the major issue is whether stability is perceived as a long-term target or just a reaction to the turmoil of the 1990s. The answer to this question determines the overall judgment of the current situation in Russia and, more importantly, the degree of the social and political consensus in the country, which is the key criterion of a truly stable development.

If it is just a reaction to the havoc and chaos of the 1990s, it seems that the country has no more than a tactical consensus allowing our society to function and guaranteeing continuity and the minimum set of rules for social communication and the legitimacy of power. Russia was characterized by exactly this type of consensus in the 1990s when people had difficulty agreeing even on the basic political procedures for legitimizing the power and the public administration of the new Russian elite.

If the answer is that stable and balanced development has already been made an integral part of our philosophy of long-term development, we can be more content to argue that Russian society is on its way to a new stage of political evolution – formation of the ontological consensus which underlies any properly functioning long-standing democracy.

One of the reasons why the idea of stability has gained wide national appeal is derived from a very low level of political segmentation and ever-growing determination to give support to any political regime that will guarantee personal freedom and a higher standard of living and economic welfare as well as sparing the time needed to understand "who is who" in politics and what the key differences between the parties are. This process and the traditional abstaining from political activities are spurred by the low level of trust towards such political bodies as Parliament and public organizations (including political parties). To make the matters worse, a significant part of Russians mainly characterized by low and medium levels of income do not believe that they can change anything by political participation at all.

It should also be pointed out that, amazingly enough, during all the years of democracy the term "opposition" itself has retained its negative meaning and connotation. Thus, there seems still to be a long way before the status of a "normal country" is reached, where the balance of power and rotation are ensured by the healthy dynamics and constant negotiations and dialogue.

The popularity of "stability" as an all-embracing concept of a country's development can be best explained by the existence of a special type of ideology, the "meta-ideology of consensus", which many scholars attribute to the existence of a special type of syncretism in conditions of barely traceable developments of a nation's political culture. The situation of political orientation is worsened by increases in both the quantity of information and the complexity of factors constituting political reality, together with the necessity to make personal political choices and provide for yourself by yourself. That is why society has generated a new set of political preferences to a great extent reflecting changes in political identity. Such notions as "stability", "order", "social justice", and "strong state" are the key terms and the core.

We are thus facing a unique situation: society is consolidated without any need to take into account the major profile of political identity of any citizen in particular. When a given society has made just the very first steps in the formation of a new identity, such consolidation is very necessary in order to allow different segments and groups to accept the existence of each other. Of course, further in this development, the question can arise of how efficient this political "consensus" is if its participants pay no attention to the methods in which the "consolidation" is achieved and "order" and "stability" are maintained: there multiple examples of such social developments where the consequences are not only economic stagnation on a national scale but also the emergence of an authoritarian regime or even its further transformation into a fascist one.

On the other hand, the very fact of such ideology formation can by itself be a sign of a completely different approach to solving various social problems: the majority of the nation seems to be satisfied with the level of state interference in their life and sees the state as a power capable of working out and implementing certain rules and laws that will serve as an essential basis for the self-actualization of individual citizens. The state is also responsible for maintaining a certain stable political and economic pattern of social development, but it is even more important to give the citizens enough space for personal freedom and business activity. This kind of agreement is rather unique in Russia's history because it seems that – almost for the first time in its history – the country has quite a substantial number of people who ask for individual freedom and self-actualization. Hence, it is surprising that in the reality of pursuing one's own welfare, political participation is perceived as an annoying formality and unwanted duty which should be minimized if possible. As a result, we face one of the most recently established concepts and phenomena in Russian political consciousness – people opt for a strong "stable" state which should also guarantee their personal space and freedom accompanied by low taxes and other favourable conditions providing growth of savings and salaries. This trend we can trace in almost all recent surveys. This belief is the perceptional core of the stability lure. The society generates this simplified demand and the state has already taken up the challenge and adjusted to the intricacies of interaction with its own people.

The traditional Russian fear of the future should also be mentioned. Regardless of the source, its current influence is revealed in the almost total unpredictability of the major patterns of political behaviour which it triggers in the people. One of the traditional patterns of this sort is the Russian indifference to various political institutions accompanied by an almost uncritical and irrational belief in the person at the head of the country.

The whole process of Russian political transit is complicated by the existence of people with absolutely different political experience and habits. One of these types can be called an "authoritarian" person inherited from the Soviet Union and another is a "democratic" person who has appeared in the course of Russia's recent development. The interaction between them is not easy, but, amazingly enough, the very basis for coexistence is provided by the concept of "stable development and balance". Nonetheless, what we see is not a sign of a real consensus in society but just a way retain a temporary balance and the status quo. Due to this superficial consensus the country managed to overcome the identity crisis of the 1990s and promote some basic values among the new generation.

Stability as a Paradigm of Development

It seems that the stability of social and economic development is one of the most efficient concepts to follow when a society is going through the initial stage of new identity formation, allowing a consolidating of society around the basic set of values and, even more important, providing some time for the mutual readjustment of different segments and social groups.

Problems arise when stability becomes the target of political and social development. In this case, society cannot be saved from the prospect of a substantial toughening of the political regime (possibly taking the shape of a police state), losing its dynamics and developing extremely low potential for meeting various challenges.

If we think of development as the process of making political power more efficient and enhancing its capabilities of solving a number of pressing issues of various nature, we have to examine the dynamics of intra-group processes. There is a close link between the efficiency of an institution and the method of maintaining the balance of a group, thus making stability of communication one of the key concepts. Even more significance can be attributed to this phenomenon if we analyze the long-term perspective of group dynamics. On the other hand, the belief that a paradigm of stable development can be stimulated in this manner has already been proved absolutely wrong. And ironically, at first it was the Soviet Union that attempted to implement this very project (the traces of this thinking can still be found in the political identity of present-day Russia). That experiment proved that such stable development does not allow flexibility in solving upcoming conflicts. The overall neglect of certain trends puts the society on the brink of a collapse in several decades' time, depriving it of internal balance.

The Russian longing for stability is a huge destabilizing factor in itself. It gives the ruling elite an illusion that they enjoy an almost unlimited national support, while in reality stability comes only from people's abstention from political participation, their dislike for opposition, and their willingness to exchange their votes for undisturbed personal freedom.

Abstention from political participation among the majority of Russians creates a situation in which the responsibility for political decisions is shifted entirely to the ruling elite. The citizens do not (and do not want to) be a part of the picture, thus taking the easiest path of indifference. This would not matter much if it did not lay the grounds for multiple conflicts in the future, being ominously reminiscent of the experience and the causes of the demise of the Soviet Union.

The Russian population is also characterized by a low level of political awareness. Most of the people go to the poll booths without any idea of the party program they are voting for and hardly knowing the differences between parties and programs. This further decreases the chances that parliamentary mechanisms will obtain a bigger role in Russia. On the other hand, there are quite a few politicians who are ready to use this situation for their own purposes; they reduce their program to a number of general principles and put forward a mix of the most popular slogans. That deprives the parties of one of their basic functions; the bringing together of the different interests of different social groups in society. This automatically postpones the solution in a number of currently latent conflicts in the society, prevents feedback and makes it necessary for authorities to interfere and regulate the situation using their "administrative" tools. The stability that is so actively promoted now is thus nothing but a temporary solution to the problems.

On the other hand, this demand for stability at all costs suits the government; it allows the creation of a passive society that is easily manipulated and greatly simplifies the task of managing it.

In that sense, stability as a concept of political development can be best compared with a game which allows the preservation of a status quo and makes political communication work but doesn't solve anything. According to the leading Russian psychologists L. Vigotsky, craving for stability occurs "in such a situation when certain unrealized and latent tendencies become visible".[6] It should also be added here that this very game follows genetically inherited patterns of communication and solving conflicts[7] and can be used effectively by the political management as a tool of restriction as well. It allows no experiments or shifts in the balance of power among the elites.

In conclusion, I would like to pose a question: is dynamic stability feasible and can it prove to be efficient in the long-term? This question can only be answered in the affirmative if the system of communication between the power and society is radically changed. The new model will need political stability as the vital component that ensures continuity, but change should also be ensured by a system that can be best called "reflexive management", based on fast response to challenges, flexibility, and reaction to the feedback coming from the grass-root level of society.

6 L. S. Vigotsky, Game and its Role in the Psychological Development of Children. Moscow: 1986, p. 183.
7 B. I. Khasan, Constructive Conflict Psychology. Moscow: 2001, p. 115.

Alexei Zudin

Regime and Rotation: What's Next?

On May 7, 2008 Vladimir Putin vacated his presidential post and transferred power to Dmitry Medvedev. The managed rotation of power was successfully completed. Two circumstances give special significance to this fact. Firstly, up to the last moment such a turn of events seemed improbable. Since the 2000's the governing elite, headed by Putin, had set up a "monocentric" regime, which was not only personalised by a particular Russian president, but was actually centred on him. It was quite difficult to imagine that the key player of the regime would leave this central position. Secondly, the withdrawal of the key player made the transformation of the established political regime inevitable. For an adequate understanding of the changes, which started after the managed rotation, it is necessary to define how they correspond to the existing concepts and in what way they were determined by the regime's development during the previous period.

The Features of a Post-Soviet Cycle

The transformation of political regime in Russia was for a long time perceived through the lens of a transition paradigm and rated by its location on a scale ranging from authoritarianism to democracy. The modification of this approach – caused by the introduction of intermediate concepts (first "democracy plus adjectives", then "authoritarianism plus adjectives")[1] and a criticism of transitological

1 David Collier and Steven Levitsky, "Democracy with Adjectives: Conceptual Innovation in Comparative Research", World Politics, 49, April 1997; Steven Levitsky and Lucan Way, "The Rise of Competitive Authoritarianism", Journal of Democracy 13 (April 2002).

approaches by Russian authors[2] – was not supported by an alternative paradigm. The situation changed after Henry Hale, an American political scientist, put forward a "cyclic model" which offered an explanation for the phenomenon of "coloured revolutions" as an alternative to transitological ones (coloured revolution – a step towards democracy). In his opinion, the transformations of post-Soviet political regimes ("elections – consolidation – fragmentation") are more likely to be defined by their institutional features than by a shift on a classical transitological scale. All post-Soviet regimes (with the exception of those in the Baltic States) with different variations were classified as being a single type, named "patronal presidency" (originally, the term "superpresidency" was used)[3], which concentrate a huge amount of formal and informal power in the hands of incumbent president. The patronal presidency creates a "winner-take-all" situation. Patronal powers impose clientelistic relations upon other players, and make key national elites, primarily regional leaders and big business, dependent upon the president.

During elections elites do everything to confirm their loyalty to the patronal president. But as soon as it is perceived that the incumbent president will be leaving his office (with the termination of the maximum term of office, or confronted with the withering of political support), elites began to defect their "patron", trying their best to guess and join the future winner. (The cost of a wrong decision is very high, as huge patronal power will, once again, be concentrated in the hands of the future president.) After the national presidential election, the basic phases of a cycle will be repeated once more: "presidential elections – patronal presidency regime consolidation – maximum term in office approximation/withering of political

2 See, for example, V. Gelman, "Post-sovetskiye politicheskiye transformatsii (Nabroski k teorii)" [Post-Soviet Political Transformations (Sketches to the Theory)], "Politiya", 2001 no. 1; V. Gelman, "Stolknoveniye s aisbergom: formirovaniye kontzeptov v izuchenii rossiiskoi politiki" [Collision with an Iceberg: Concept Formation in the Study of Russian Politics], Polis, 2001 no. 6.

3 H. E. Hale, "Russia's Presidential Election and the Fate of Democracy: Taking the Cake" (electronic version: http://www.fas.harvard.edu /~ aaass/newsnet/2004-05-cover.html).

support - defection of elites - fragmentation and disintegration of regime - new presidential elections etc."[4]

The cyclic model provides an opportunity to reinterpret the development of post-Soviet political regimes. The particular value of this concept is the separation of institutional dynamics from the processes of "transition". The premises upon which the model in question is based are no less important. Political transformation is classified as a long-term process with a potential to develop in different directions. In this sense Hale returns to the foundations of a transitological paradigm, lately modified by enthusiasts of accelerated democratisation, that often tend to produce inconsistent and unstable outcomes. (The most obvious examples are Georgia and Kyrghyzstan. The transformational processes in Ukraine after the orange revolution are far from being over and cannot be defined with the required certainty.)

But Hale's hypothesis concerning the development of the electoral cycle of 2007–2008 in Russia and its political outcomes proved incorrect. In spite of the instances of the successful "managed" rotations in Russia in 2000 and Azerbaijan in 2003, Hale rated the ability of the successor of the outgoing president to win elections as being highly restricted. In his opinion, even a mild uncertainty about future access to resources of the patronal presidency would push elites to forge a counter-coalition against the successor. It was assumed that if Putin decided not to run for the third term, the parliamentary elections of 2007 would be more similar to those of 1999 than of 2003. And if the elections of 2007 would not bring high certainty about the winning candidate to presidency, the presidential elections of 2008 would become more competitive, with a real prospect of a transfer of power to the opposition for the first time in post-Soviet history. The author assumed that Putin could make

4 H. E. Hale, "Institutions and Transitions. Democratization, Autocratization, or No '-ization' at All in the Former Soviet Union?", prepared for the 2005 Annual Meeting of the American Political Science Association, September 1–4, 2005 (especially pp. 6–10); H. E. Hale, "Regime Cycles. Democracy, Autocracy, and Revolution in Post-Soviet Eurasia", World Politics 58 (October 2005), pp. 133–165; H. E. Hale, "Democracy or Autocracy on the March? The Colored Revolutions as Normal Dynamics of Patronal Presidentialism", Communist and Post-Communist Studies, vol. 39, no. 3, 2006, pp. 305–329).

an attempt to break the "patronal cycle" to ensure his successor's victory, but was sceptical about the success of such an endeavour.[5]

It seems that the flaws of Hale's concept are caused by the limits of a purely institutional approach when applied to the dynamics of post-Soviet regimes and the underestimated importance of politics – the strategies of the actors and the impact of context. The author distinguishes the role of public opinion and other noninstitutional factors in the dynamics of post-Soviet regimes. (Among the reasons for the defection of the elites not only the expiration of maximum term in office is specified, but also the withering of popularity of the incumbent president as a result of a crisis or a military defeat.) But it seems that the necessary conclusions were not drawn.[6] Hale admits that strong presidents are able to subordinate elites, prevent their defection on the eve of an election campaign, and from time to time arrange electoral victories of their successor. But even the successor's success cannot essentially change anything. Hale does not make a distinction between the political effects of Putin's possible re-election for the third term and a managed power rotation: in both cases, the most probable outcome is the reproduction of an

5 H. E. Hale, Russia's Presidential Election and the Fate of Democracy: Taking the Cake (electronic version available at: http://www.fas.harvard.edu/~ aaass/newsnet/2004-05-cover.html).

6 H. E. Hale, "Institutions and Transitions. Democratization, Autocratization, or No '-ization' at All in the Former Soviet Union?", pp. 2, 10, 23. In the latest publications the author pays attention to the autonomy of the public opinion in Russia's political process and also to the analysis of close relations which have developed between the president V. Putin and a large segment of voters in 2000–2007. See T. J. Colton and H. E. Hale, "What is the Putin Factor? Explaining Electoral Support in Hybrid Regimes", prepared for the Annual Meeting of the Midwest Political Science Association, Chicago, April 3–6, 2008; T. J. Colton and H. E. Hale, "The Putin Vote: The Demand Side of Hybrid Regime Politics", prepared for the conference "The Frontiers of Political Economics", Higher School of Economics, Moscow, May 29–31, 2008. The important part of the new research program which, according to Hale, attempted to compensate shortcomings of the former approach which too closely adhered to the concept of transition, should be a study of the factor of political leadership (see: H. E. Hale, Institutions and Transitions … p. 23). It is possible that the completion of the theory of a cyclic model, in which the author is now engaged, will help increase its explaining power, but this will demand the specification of some concept positions, at least in its present variant.

"autocratic cycle". Putin's successor will create his own monocentric regime. The established institutions keep the patronal presidency in a cyclic trap.

Meanwhile, available experience allows us to qualify the patronal presidency as being a more complex phenomenon which is conditioned by institutional as well as by noninstitutional factors, and not only in descending phases of a transformational cycle (circumstances inducing elite disloyalty) but also in ascending phases (elections). Patronal power does not proceed automatically from the extensive formal and informal prerogatives of a nationally elected president. More adequately it could be described as a potential which can be realised by the president under certain circumstances and in a number of ways. The political power acquired by the president at elections plays a crucial role in obtaining patronal resources: the institutional component of power can only be legitimised by an electoral victory. But besides electoral success, the political context of winning elections is important, i.e. the scale of electoral support and the political conditions of victory – for example, whether it was achieved in one or in two rounds. The effects of patronal presidency are shown to their full extent when institutional and political components complement each other.

However, there is also a possibility of their mismatch. For instance, the president could be constrained by a set of 'selective' (individualised) agreements with loyal national elites, especially those created by market and democratic reforms (as Boris Yeltsin was in 1991–1993 and in 1996–1998). In these circumstances the president is not able to use the full potential of his patronal power.

Moreover, the patronal power of the president in post-Soviet Russia happened to be limited not only from the outside (by 'selective' agreements with elites), but also from the inside. Despite broad formal prerogatives of the president (epitomised in the phenomenon of "superpresidency"), the institutional order is not coincident with a "pure" presidential system: the president is not the head of the government, and the candidacy of the prime minister must be approved by the State Duma. The recurrent strengthening of the prime minister is a political fact that has occurred whenever a prime minister has held his office for a long period. The growth of Mikhail Kasyanov's political autonomy in 2002–2003 (in the shadow of the politically powerful president Putin) proves this fact to an even larger degree than the rise of the political strength of Viktor Chernomyrdin next to the failing Yeltsin in 1997–1998. (The political build-up of prime minister Evgeni Primakov at the end of 1998 and the beginning of 1999 was more complex case, conditioned not only by the internal dynamics of the presidential power, but also by the political intervention of the State Duma.) But when the accumulation of political autonomy by the head of the government was matched by the political decay of the president (as happened in the case of Chernomyrdin and Yeltsin), the heavyweight prime minister found himself as the distributor of patronal resources, nominally controlled by the Russian president.

Hale does not differentiate Putin's political regime from that of Yeltsin, considering both an embodiment of the patronal presidency. But a coherent application of his approach ends with a paradox: under President Yeltsin patronal power was barely evident (Hale cautiously recognises that the given circumstance represents "a difficult case") while the patronal model of institutional dynamics manifest itself in a different way during Putin's governance.

The Making and Dynamics of a Monocentric Regime

Under Putin, the transformation of the political regime into a monocentric one created the conditions necessary for overcoming the cyclic trajectory. The making of a new regime was accompanied by a status coup at the higher levels of power: formerly peripheral elite factions (the "petersburgians" and "siloviki" - proponents of a tough approach, connected to security, military and paramilitary government agencies) moved to key positions in the power system, while oligarchs and the regional governors lost their privileged positions. For a long time the changed composition of the higher circles framed the perception of Putin's regime by the bulk of political observers as well as academic researchers in Russian politics. (A good example is the concept of "militocracy" that was advanced by Russian sociologist Olga Kryshtanovskaya.)

But it was the political context of the presidential elections in March 2000 which played the decisive role in the acquisition of new political qualities by the political regime: the high personal popularity of the new president and his victory in the first round of elections. The single round of elections changed the relations within the triangle "president - elites - mass voters". For the first time since the beginning of the 1990s, the president managed to restore direct links with electoral majority. This was made possible by the high popular support of the new president and by the fact that the elites lost their critical role as electoral intermediates. The role of public opinion in the regime's political base was substantially increased. Along with a set of "selective" agreements with national elites, the presidential power appeared to enter into a new kind of agreement - an implicit social contract - with mass voters. New political resources that helped restore the political strength of presidential power were acquired (the "Putin majority").

Unlike Boris Yeltsin, who was not able to fulfil the patronal potential of presidential power due to contextual factors, Vladimir Putin really became a patronal president. But there was another important difference: his power was based not only on institutions but also on popular political support, and the importance of this second pillar of presidential power gradually began to grow. The president became the strongest political player and presidential power no longer needed elites as critical intermediaries to public opinion and mass voters. The "parallel" contract with the "masses" changed the conditions of "selective" agreements with elites in favour of the presidency. From that time it was possible to impose a sort of consensus upon the elites. The terms of the imposed consensus implied that the elites were subordinated to the new governing group, that they abandoned temptations to challenge the presidential power at national elections, and also that they accepted the loss of control over some of the resources which they had managed to acquire during the second half of the 1990s. But the underlying motivation was not the restoration of "selective" agreements based upon the early patterns of 1991–1993. Capitalising on new resources, the president obtained the opportunity to reshape his relations to the elites from a position of strength. An important contextual meaning in the relations of presidential power with the strongest elites after 2000 elections was the "winner-loser" frame. It was well remembered that politically binding relationships were broken by the strongest elite factions, who broke with the politically weak Russian president after the August default of 1998. They went into political opposition, built two coalitions and challenged the political regime at the 1999 Duma elections, but ultimately failed.

After his victory in the presidential elections of 2000, Vladimir Putin as Yeltsin's successor was not obliged in political and moral terms to restore the terms of "selective" agreements which defected elites had broken after August 1998.[7] The new position of political strength and the "winner-loser" frame enabled the presidential power to change the direction of political development: reforms during Putin's first term in office were driven by the general idea of political centralisation (the introduction of federal districts, restructuring of the Federation Council, and transformation of the federal political parties into central players of the electoral process). Regional leaders and oligarchs lost institutional and political means to put pressures upon the presidential power.

The politically empowered president played the leading role in the reorganisation of relations with the elites. But the transformation of relations between the federal executive and the State Duma, which had become the main political problem for Boris Yeltsin, was provided by the new "party of power" - United Russia. From the very beginning United Russia differed from the old party of power (Our Home - Russia) by its higher degree of political centralisation and efficiency. The increased political capacity of the new party of power made the unprecedented success of the reform program possible through its legislative promotion. Nevertheless, during the first period of Vladimir Putin's governance, the elites retained autonomy in their institutional domains. Having lost critical influence on federal institutions, governors preserved their personal

7 A. Zudin and A. Ryabov, "Osobennosti kampanii-2000 i ih vliyaniye na konfiguratsiyu vlavstvuischei elity", Rossiya v izbiratel'nom tsikle 1999–2000 godov. Pod red. M. Makfola, N. Petrova I Ryabova. M., Moskovskii tsentr Karnegi [The Properties of the 2000 Presidential Electoral Campaign and Their Influence upon the Configuration of the Ruling Elite, in: Russia in the 1999–2000 Electoral Cycle, ed. by M. Makfola, N. Petrov and A. Ryabov. Moscow Carnegie Centre], 2000, pp. 421- 428.

electoral machines (having redesigned them into local branches of United Russia) and political control over respective territories. Unlike presidential elections (after 2000), the dependence of the results of State Duma elections upon the political resources of regional elites remained. For that reasons, the initial course of the Kremlin, aimed at disempowering local bosses on the eve of the parliamentary elections of 2003, was replaced by a policy of co-optation: the governors were made "political locomotives" in the party lists of United Russia.[8]

The consolidation of the monocentric regime was the main political result of the electoral cycle of 2003–2004. The governing group became more homogeneous. Petersburgians consolidated their grasp at the centre of the governance system, while the "family" group, its key members being the inner circle of relatives and friends of the former president Boris Yeltsin, lost its positions. Vladimir Putin acquired the political strength of the dominant player and the monocentric regime became openly personalistic. United Russia assumed a hegemonic position in the State Duma (a "constitutional majority"), while the factions of socially advanced groups, associated with the liberal parties (the Union of the Right Forces and Yabloko) lost parliamentary representation. The political enhancement of the leading party was accompanied by the transformation of the party system to a "one and a half" model.

The role of political parties began to change, but the trend was multidirectional. The functions of political parties were still

8 For more on the main political trends at the time of 'monocentric' regime formation, see: A. Zudin, "Rejim V. Putina: kontyri novoi polititcheskoi sistemy", Moskovskii tsentr Karnegi [The Political Regime of V. Putin: An Outline of the New Political System, Moscow Carnegie Centre], April 2002; A. V. Lichtenshtein, "'Partii vlasti': elektoral'nuye strategii rossiiskih elit nakanoune viborov" [The "Parties of Power": Russian Elites' Electoral Strategies at the Eve of Elections]; V. Gelman, G. V. Golosov, E. Yu. Meleshkina, "Vtoroi elektoral'nyi tsikl v Rossii" [The Second Electoral Cycle in Russia], 1999–2000 gody. M., Ves Mir, 2002; R. Smyth, "Building State Capacity from the Inside Out: Parties of Power and the Success of the President's Reform Agenda in Russia", Politics and Society, Vol. 30 No. 4, December 2002, pp. 555–578; O. J. Reuter, "Parties of Power as Institutions: Commitment Problems and the Rise of United Russia", Prepared for presentation at the Annual Meeting of the Midwest Political Science Association, April 12–15, 2007

restricted to elections and the legislature. Their autonomy decreased but, due to the partial "partisation" of the regional parliaments, their presence in the legislative process strongly increased. The political atrophy of the parties was paradoxically combined with their institutional strengthening. The political opposition was disempowered and marginalised,[9] and federal elections began to acquire the properties of plebiscite.

The main political reforms during the second term of Putin's governance (the abolition of direct elections of governors, the introduction of a proportional electoral system to the State Duma, and a mixed electoral system for regional parliaments) dramatically changed the balance of power between the presidency and regional leaders. The semi-proportional election system unlocked regional political regimes and made them vulnerable for control from the federal centre. The abolition of direct elections and reduction in the autonomy of institutional domains made party membership in United Russia a necessity that regional leaders could not afford to reject.[10] Further steps to marginalise political conflicts were made: some functions of political representation were actually transferred from the federal parliament to the Public Chamber, a body with consultative functions that was created in 2004. Roughly from that time political governance was partly supplanted by administrative governance, a process accompanied by the bureaucratisation of elites and a replacement of public actors by agents of the administrative system. (One of the manifestations of the process was the establishment of public advisory councils before the federal ministries and agencies since 2005).

The political build-up of the dominant player was accompanied by the revision of relations with big business (as epitomised in the case of YUKOS). The private sector's position in the Russian economy became somewhat ambivalent and big business lost its partner status in its relations with the government: its role was confined to being the supplier of resources while the functions of strategic goal

9 V. Gelman "Political Opposition in Russia: A Dying Species?", Post-Soviet Affairs, 2005, vol. 21, no. 3, pp. 226–246.

10 O. J. Reuter, "Parties of Power as Institutions: Commitment Problems and the Rise of United Russia", prepared for presentation at the Annual Meeting of the Midwest Political Science Association, April 12–15, 2007.

formation were assigned exclusively to the governing elite, speaking on behalf of the state. The institutional form of government-to-business relations shifted from the "regime of consultations" model, practiced from 2000 to 2003, to a system which conventionally could be called "administrative guidance". The government started to perceive itself as the leading force of the Russian economy. After the establishment of the "institutes for development" (essentially a group of state corporations) a shift in economic policy to some form of "competition state" became evident.[11]

For the dominant player, the importance of "direct" social contract with mass voters increased (especially with the start-up of socially-oriented "national projects") and the political relevance of "selective" agreements with elites decreased. The institutional model of relations between the presidential power and national elites also began to change: a set of individual contracts began to to be supplanted by generalised and collective ones with the prerequisite of membership in the political party United Russia. The transition from personalised and particularistic contracts to the universal institutional agreement was facilitated by the crisis of the first generation of post-Soviet elites: their propensity for collective action, political leadership and innovation in functional domains started to decrease.[12] The monocentric political regime was consolidated on a conservative basis and its degeneration towards authoritarianism seemed inevitable.

11 A. Yakovlev, "The Evolution of Business – State Interaction in Russia: From State Capture to Business Capture?", Europe-Asia Studies Vol. 58, No. 7, November 2006, 1033–1056; A. Zudin, "Gosudarstvo i biznes v Rossii: evoluytziya vzaimootnoshenyi" [Tha State and Business in Russia; the Evolution of Relationships], Neprikosnonennyi zapas, N6 (50) 2006, pp. 200–212.

12 Z. Zudin, "Rossiiskaya elita na rubeje pokolinii", Rossiya: vchera, segodnya, zavtra. S tochki zreniya ekspertov". M., TsPT – Assotsiatsiya menedjerov, 2007, pp. 82–94 [Russian Elites on a Boundary of Generations", in: Russia: Yesterday, Today, Tomorrow. From the Point of View of Experts. M. CPT, Association of Managers, 2007, pp. 82–94].

The Dominant Player as a Source of New Political Dynamics

The role of political factors in regime development increased after a dominant player appeared on the scene. The dominant president became increasingly invulnerable to the pressures of the institutional cycle and emerged as an independent source of political dynamics. And while the functioning of a political regime basically corresponded to the logic of a patronal president in the time period from 2000 to 2005, afterwards a new factor became evident: the strategy of a dominant player.

The purpose was not to remain in office, but to control the managed rotation of presidential power. For the first time this purpose was publicly declared by Vladimir Putin in the summer of 2003. The main task of the dominant player became a fight against institutional and political inertia, and to achieve this task anti-cyclical strategies were employed. From that moment, the movement of a political cycle ceased to be inertial and each subsequent phase became managed.

The Managed Fragmentation of a Regime

During the preparation for the electoral cycle of 2007–2008, the strategies of the dominant player enabled him to overcome two inertial scenarios: the natural outcome of the post-Soviet political cycle (in case the incumbent president simply would leave the office with or without the promotion of a "successor"), and the transformation to a personalistic authoritarian regime (in case the incumbent president would make an unconstitutional decision in favour of the third term). In order to build a strategy capable of overcoming the logic and neutralising the effects of a post-Soviet patronal presidency, it was necessary to solve a number of serious problems. The prospect of leaving the office inevitably transforms the president into a "lame duck" with decreasing political support by elites and public opinion. Thus the lame duck prospect was identified as being the main peril to the managed rotation. The main provision for a successful managed rotation of power was the retention of political strength by the president in office. The choice was made in favour of keeping a state of high political uncertainty as the key strategy.[13] Subsequent events showed that the "strategy of uncertainty" (accompanied by high political activity of the dominant player) was able to solve the lame duck problem: the leaving president managed to retain political strength and the loyalty of elites during the entire period.

The state of high political uncertainty was kept by two principal means. The first of these was the air of unreality around the prospect of power rotation: until the last moment it was not clear whether the publicly promised rotation would really take place despite Putin's periodic statements that he would not violate the constitutional provisions. A political campaign in favour of a third term for Putin, carried out by the prominent members of elite and the loyal part of mass media, became an important part of the strategy of uncertainty. In fact, public endorsement for the third term option was integrated in the informal ritual of political loyalty to the incumbent president. (The leaders of the political opposition became involuntary participants of this campaign: they perceived it as practical proof of Putin's hidden autocratic intentions.)[14]

The second option was to keep the uncertainty about the candidacy of the prospective "successor". Two alternative candidates with comparable status positions were chosen for this role – vice-premiers Dmitri Medvedev and Sergei Ivanov. After the attention of the media was concentrated on them, the "virtual" competition between the contenders for the role of successor began, and public opinion began to differentiate and mobilise behind the competitors.

The logic behind the dynamics of the virtual rivalry was also subordinated to the main strategic goal – to maintain this political uncertainty as long as possible. At the beginning it was Sergei Ivanov who possessed more political weight, and later Dmitri Medvedev

13 The hypothesis that the leaving President will use the strategy of uncertainty in order to preserve his political strength was put forward by the author in the middle of 2006. See A. Zudin, "Rossii bol'she ne nujna odnorazovaya 'gosudarstvennost'" [Russia Doesn't Need 'Disposable' Statehood Any More], Politcom.ru, 29/06/2006.

14 A. Zudin, "Tretii srok – igra, kotoraya nujna vsem" [The Third Term: A Game that Suits Averybody], http://www.politcom.ru/article.php?id=1908

caught up with him. After the latter was started to be perceived as the most probable successor and political uncertainty began to decrease, their fortunes changed once more and Sergei Ivanov's public activity and popularity began to rise again. According to the opinion poll of the Levada centre, in March 2007 Sergei Ivanov overtook the contender in electoral support in the second round of simulated presidential elections. In April the gap even increased, but in May an approximate parity between the competitors was restored. Meanwhile the pronounced differences in the contenders' images began to fade, and both of them acquired the multidimensional qualities (identical compatibility with liberals and siloviki) which only the incumbent president used to have.[15]

By the end of the summer of 2007 the rivalry between Medvedev and Ivanov was politically exhausted. The attempt to prolong the state of uncertainty by bringing an unknown third candidate into the game (Igor Shuvalov's statement in second half of June 2007) was not successful. Presumably the elites had tired of associating the situational advantage of a contender at a given moment with the final outcome of the contest. At the end of August 2007, manifestations of a spontaneous (or not so spontaneous) consolidation around Sergey Ivanov, who once more achieved a slight lead in public opinion, jeopardised the managed rotation with the real prospect of reducing Vladimir Putin to a lame duck. Victor Zubkov's appointment to prime minister at the beginning of September 2007 prevented the untimely consolidation of elites around Sergey Ivanov. The possible third candidate to the role of "successor" was advanced in a fashion that was hard to ignore: the importance of the position and the precedent that had been established by Yeltsin and Putin in 1999 meant that the holder of the office of prime minister was the natural candidate to succeed the president. Practically immediately the new prime minister publicly declared that he did not exclude the option to become a candidate in the upcoming presidential elections. That was unusual for a high-ranking official who had to avoid personal statements on electoral topics.

15 "Zakat preyemnikov" [The Fall of Contenders], Moskovskiye novosti, 2007, no. 21, 01/06/2007.

As a candidate for the office of president Victor Zubkov quickly outstripped both Sergei Ivanov and Dmitri Medvedev by a wide margin. Political uncertainty was restored – along with the managed character of preparation for the rotation of power.

The strategy of uncertainty was combined with relevant activities in other fields. Information was distributed of the inevitable political fragmentation of the regime, and also of the will to restrict and control this process.

The transformation of the regime had actually began before the electoral campaign of 2007–2008. The new configuration at the higher circles of power was created earlier in order to facilitate the players' subsequent adaptation to it. Aside from the reproduction of political uncertainty, the introduction of competing candidates to the role of successor corresponded to the logic of managed fragmentation of the political regime. Two more important steps also were made along "managed fragmentation" model. The most pronounced was the demonopolisation of the Kremlin party niche: Fair Russia, the second political party, supported by the incumbent president, was established next to United Russia in 2006. The trend towards the monopolisation of the political market, prevalent during the previous period, was stopped.[16]

The other manifestation of the managed fragmentation of the political regime was the establishment of six new public corporations, controlled by the federal government. The list of corporation, created from May until November 2007 included: Vneshekonombank [a state "Bank for Development and Foreign Economic Affairs"], Rosnanotech [Russian Corporation of Nanotechnologies (RUSNANO)], FSR JKH (Fond sodeistviya reformirovaniyu jilishno-kommunal'nogo hozyaistva) [a state "Foundation for Housing and Utilities Reform

16 A. Zudin, "Predvybornaya kompaniya 2007–2008 godov: klychevaya intriga i stzenarii transformatzii politicheskogo rejima" [The Pre-election Campaign 2007–2008: The Central Intrigue and Scenarios of Political Regime Transformation], Politiya, no. 2, 2006; V. Gelman. "Transformatsiya rossiiskoi partiinoi systemy: monopolizatsiya politicheskogo rynka" [The Transformation of Russian Party System: The Monopolisation of Political Market], Obshestvennyie nauki i sovremennost, no. 1, 2006; A. Zudin, "Novaya 'Zhizn' v polutora-partiinoi systemye" [New "Life" in the One-and-a-half Party System], Politcom.ru, 10/9/2006; A. Zudin, "Privivka ot 'partii vlasti'" [Vaccination from the "Party of Power"], Moskovskiye Novosti no. 5, 09/02/2007.

Assistance"], GK Olimpstroi (Gosudarstvennaya Korporatziya po stroitel'stvu olimpiiskih ob'ektov i razvitiyu goroda Sochi kak gjrnoklimaticheskogo kurorta) [a state corporation for the construction of Olympic Winter Games facilities and the development of the city of Sochi as a mountain climatic resort], Rosatom [the state atomic energy corporation], Rostechnologii [a state corporation "Russian Technologies"]. The establishment of new state corporations is usually treated as an indicator of a turn to "dirigisme" in economic policy. But it also had a political dimension: economic decision-making was partly removed from the functional domain of the federal administrative elite and was delegated to semi-independent centres – the top management of the state corporations.

Top managers at some of the new state corporations were also closely connected to Vladimir Putin. (The most frequently mentioned examples are Sergei Chemezov in Rostechnologii and Mikhail Kovalchuk in Rosnanotech.) The leaving president used his patronal power to link the future prime minister to the administration of resources that formerly were concentrated around presidential power. The managed fragmentation of the regime before and during the election campaign of 2007–2008 created the preconditions for deconcentration at the centre of the political governance system.

Deconcentration in Higher Circles

An important role in the preparation of the managed rotation was played by "castling", i.e. the shift to new positions in the system of governance. Castling happened to be a complex process and was carried out over a long period of time. The transition of the leaving president to the new political shape began half a year before he left office, and was accomplished in two successive steps. At first a conversion of political strength was conducted: on 1 October 2007 Putin accepted the offer by the leadership of United Russia to head the party electoral list. He also publicly displayed an interest to head the government, with some preconditions: United Russia would have to win the elections to the Duma, and the newly elected president would have to be a deserving figure "capable of working in a team".

In this manner, favourable conditions for the conversion of political power were created: after the parliamentary elections the political strength of the leaving president would be reconfigured in a new and legitimate fashion, not related in formal terms to presidential power. In any case, Putin managed to jump on a departing train: if he hadn't headed the list of United Russia then he would have turned into the hostage of the winners of the forthcoming parliamentary and presidential elections. The outcome of electoral cycle of 2007–2008 would vest United Russia's party leadership and the new Russian president (no matter who it would be) with a fresh electoral mandate. After new political leaders had been chosen by the voters, the main source of Putin's political strength would inevitably have began to melt away: public opinion ratings acquire real political strength only when they are based on the legitimately expressed political will of voters. In any case, the chances of the former president to retain public support once he left the centre of political governance would have been greatly reduced. Despite the sketchy similarity to the previous electoral cycle, the parliamentary phase of elections of 2007–2008 was subjected not to another round of consolidation of a monocentric regime, but to the alternative logic of managed deconcentration.

Putin's direct involvement in the Duma election campaign bred important political effects. Regime fragmentation was suspended. In full conformity with the cyclic model, the prospect of power rotation was capable not only to turn the leaving president into a "lame duck", but also to abolish the entire system, that was built upon him. The disintegration of the political regime could have begun during the electoral campaign of 2007. In September, two trends became visible: the electoral ratings of United Russia dropped as the electoral mobilisation of other participants started. The prospect of the negative mobilisation along the line "all versus the leading party" came out. The FOM ("Public opinion" Foundation) poll of 5–7 December 2007 discovered a considerable increase in negative motivations among electoral supporters of the Communist Party and Liberal-Democratic Party as compared with 2003.[17]

17 FOM, "Motivy i faktori elektoral'nogo vybora. Otchet. Opros naselenyia" ["Public Opinion" Foundation. Motives and Factors of Electoral Choice. A

Under these circumstances, it seemed that the traditional political blessing from Vladimir Putin might not be enough to provide United Russia with the necessary electoral results: after all the president (no matter how popular) was leaving his office while the party would remain in the State Duma long after his departure. This prospect disappeared after Putin became personally involved in the parliamentary campaign. The electoral attractiveness and mobilisation capacities of United Russia increased considerably, but the ability of the party to appropriate the change was sharply reduced: the leaving president simply crushed the party with his political weight. Putin's intervention provoked another political effect. The personification of elections to the State Duma (at least, for supporters of United Russia) transformed the voting for a political party into a plebiscite on the political future of the leaving president.

Political uncertainty also began to decrease, but in an uneven way: at first Putin's new role in the system of governance was established, and only then the personality of the "successor" became certain. The reduction of political uncertainty brought into being a new set of political expectations, with a disciplinary impact upon the governing group and elites. But new problems were also born. The retention of a strong political role for the leaving president made it necessary to built a new balance at the centre of the governance system, presumably on the basis of a comparison of Putin's former electoral rating (as a president) with that in his new position (the electoral rating of United Russia in December 2007). The latter figure needed to be similar to the electoral result that the new president was likely to achieve at the elections in March 2008.

The change of strategy by the dominant player in the midst of election campaign also generated new political risks: the election of a two-party Duma; Putin's loss of his all encompassing status as a result of an excessive identification with United Russia; a gradual erosion of his popularity as a result of his increased responsibility for government policies; the appearance of the "technical" (inadequate) successor (in comparison with Putin practically any figure would not look impressive – the departing president could be suspected

Report. The Poll of the General Public]: http://bd.fom.ru/report/map/projects/dominant/dom0749/d074922.

of deliberately choosing a "weak" candidate); and the emergence of a conflict between the president with wide constitutional powers and a politically strong prime minister.[18] Later some of these risks managed to be neutralised by the preservation of a nonmembership status for Putin at the head of United Russia and by the transformation of the structure of government; it seemed, however, that other political risks - the dangers of a "weak" successor and the possibility of conflict at the centre of political governance - were not solved but rather postponed.

The risk of political conflict between the president and the prime minister as the result of the managed rotation was clearly perceived by the public. According to VTSIOM (a public opinion research agency), at the end of October 2007, 52% of those questioned considered the likelihood of a redistribution of power between the president and the prime minister to be inevitable if Putin became prime minister. Should such a conflict occur, voters expected that the new president - whose name they did not know at that time - would win it. (51% of the respondents believed that basic powers would be concentrated in hands of the new head of the state.) During the election campaign this kind of political risk was partially neutralised by reassurances that the governance system would remain within the constitutional framework and that no institutional changes would be made. Vladimir Putin several times publicly rejected the prospect of redistributing the formal power of the executive branch; earlier had also rejected the expediency and feasibility of a new parliamentary model in Russia. (Later, Dmitri Medvedev, who had not yet become a "successor", made similar statements.)

In the final stages of the electoral campaign the main strategic objective of the departing president was to bring the level of United Russia's electoral support closer to his presidential rating. The personification of elections to the State Duma was strengthened. In the middle of October 2007, Putin acted unexpectedly: he publicly distanced himself slightly from United Russia, whose party list he had recently headed. The extension of Putin's political appeal

18 A. Zudin, "Peredumal li president?" [Has the President Changed his Mind?], Moskovskiye novosti, no. 49 (14–20/12/2007); "Tandem ili treugol'nik?" [The Tandem or a Triangle?], Moskovskiye novosti, no. 50, 21–31/12/2007.

beyond party borders was also increased by the activity of his non-party supporters, who held a highly publicised forum in Tver in the middle of November. At the end of November 2007 the electoral campaign of United Russia's was consolidated with a "non-partisan" campaign. At a joint forum in Luzhniki (Moscow), Putin threw all his political weight in favour of electoral mobilisation around United Russia: he reminded of the linkage between parliamentary and presidential elections and declared that the well-being of the country was at stake.

Putin's public consent to head the government at the party congress of United Russia on 17 December 2007 could be considered as being the last step in the process of political strength conversion. The perspective of obtaining the office of prime minister made the departing president safe to the risk of public disassociation from the attributes of power, which seemed to be crucial for the retention of political strength. The foundation for a new political balance at the centre of the governance system was also established. The coherence of the governing group along the lines, built after 2004, played a key role for that purpose. The managed rotation made the continuity of the political course possible, but at the same time generated a new problem, the so-called "successor dilemma".[19] The successor to the leaving president had to be chosen from one of the several competing groups. No matter who would be chosen, the other groups would most likely have been dissatisfied, and this could pose an obstacle for the consolidation of the political regime around the new president.

The effects of the patronal presidency in the winner-take-all situation aggravate anxiety in groups which do not perceive the new president as being "theirs". In 1999–2000 the solution for the successor dilemma was provided by the coalition structure of the new governing group (the alliance between "family" and petersburgians). In 2007–2008 the possible effects of the successor dilemma were neutralised in another way, namely by the decision of the ex-president to share power with the new president. The possibility that Putin would remain in politics (and in power) even after he

19 Henry E. Hale, "Democracy or Autocracy on the March? The Colored Revolutions as Normal Dynamics of Patronal Presidentialism", Communist and Post-Communist Studies, vol. 39, 3, 2006, pp. 305–329.

left his office was not considered by Henry Hale in his cyclic model. Meanwhile, such a prospect ("rotation with castling") looked as the most probable. The managed rotation with the retention of the ex-president as a powerful player logically led not to the reproduction of a monocentric regime with the new president on top, but rather to a deconcentration at the centre of political governance. (The only thing which remained unclear was the position to which Putin would shift, since each of the discussed options had advantages and disadvantages.)

During the parliamentary campaign the intrigue concerning the personality of successor returned and did not disappear immediately after the announcement of the elections results. Public opinion ratings of contenders began to level off: the indisputable "virtual" leader, Victor Zubkov, saw his presence in the electronic mass-media reduced while the first vice-premiers Sergey Ivanov and Dmitry Medvedev continued to appear regularly on television. Dmitry Medvedev's image was strengthened: he conducted a series of meetings with the leaders of key religious confessions (the Russian Orthodox Church, Moslems, Buddhists and Judaists). Such public contacts had earlier been the exclusive prerogative of Vladimir Putin, and Sergey Ivanov did not obtain such an opportunity.[20] There was also an additional intrigue: how the successor would be nominated.

The nominating procedure appeared to be complicated enough. Contrary to expectations, United Russia did not receive a monopoly over the nomination. Actually the candidate for president was put forward twice. Dmitry Medvedev was nominated as the candidate of the four "pro-regime" political parties (United Russia, Fair Russia, the Agrarian party and a tiny liberal group, called "The Civil Force") at a meeting of the leaders of these parties with Vladimir Putin on 10 December 2007. The distancing of the departing president from the nomination procedure at this stage (at a meeting with party leaders, Putin only supported the decision made by them earlier) and the participation of several political parties presumably legitimised the transformation of a "successor" into a real candidate, that is, a representative of the political parties that nominated him to the office".

20 A. Zudin, "Rojdeniye vtoroi sily" [The Birth of 'The Second Force"], Moskovskiye novosti, no. 48, 7–13/12/2007.

(The special role of United Russia in this procedure was apparent only in the fact that the four parties' nomination of Medvedev was offered to Putin by United Russia's leader Boris Gryzlov on behalf of all.)

The nomination of Dmitry Medvedev fundamentally changed the political situation. The strategy of uncertainty, the chief instrument to maintain the political balance, completely exhausted itself. But the necessity to keep control over the electoral situation remained, taking into account the political weakness of Medvedev as a candidate. Probably for this reason, the political resources that the leaving president could use to resist any manifestation of a "cyclic model" reached their critical magnitude during the time between the start of presidential electoral campaign and the moment of the formal transfer of power to the new president (December 2007 to May 2008). During this time the dominant player concentrated the maximum political strength: his remaining presidential powers, his exclusively high public support, his leadership in the dominant political party which had recently acquired the constitutional majority in the State Duma, and, finally, his public declaration that he would head the government. From the start of the presidential electoral campaign Medvedev's progress as a candidate to the office was intimately connected to Putin's prospect to become prime minister.

The Medvedev–Putin tandem displaced uncertainty as a focal point of the "anticyclic" strategy. The particular qualities of the nomination process assigned clear obligations to the key presidential candidate: Medvedev was nominated not simply as a "deserving" candidate, but also as a person ready to work in close collaboration ("in tandem") with Putin. The day after his first (technical) nomination Dmitry Medvedev publicly proposed that the departing president should head the Russian government after the presidential elections. The second (public) nomination of Medvedev took place at the party congress of United Russia on 17 December. This time Vladimir Putin and other leaders of United Russia played the central role in the process. The image of a tandem received more reinforcement: the leaving president gave public consent to head the government if Dmitry Medvedev is elected to the position of president.

The circumstances of Medvedev's election on 2 March 2008 also strengthened the tandem: the possibility of direct personal links to voters was practically excluded and, hence, the prospect for the new president to become a contender of the strongest political player – Vladimir Putin. The electoral victory was won in one round, but Medvedev in 2008 – in contrast with Putin in 1999 – was not able to appropriate his electoral support: the presence of the strong departing president did not allow this. At election time the new president was politically highly dependent upon the old president: Dmitry Medvedev was primarily supported by resources that are controlled by Vladimir Putin. Anti-cyclic strategies, aimed to prevent the disintegration of a political regime in the process of power rotation, changed the meaning of voting at the presidential elections in March, 2008: not only the new president was elected (even though with resource assistance of the leaving president), but also a plebiscitarian support for the 'joint government' of the new president Medvedev and the future prime minister Putin was built.

A reorientation of the new political course, corresponding to the requirements of a joint government by Medvedev and Putin, also started at the time of the presidential campaign. It seems that the general framework of the new course was first designed by members of "Putin's team", and then worked out in detail by initiatives presented alternately by the future prime minister Putin and the presidential candidate Medvedev from the end of December 2007. The new political course was the mixture of a moderate version of dirigisme as well as liberal approaches: an innovative economy, the restricted role of the state – though the priority of the "grand projects" was preserved –, new incentives for small and big business development, the start-up of "systemic" anti-corruption policies, the reinforcement of the legal foundations of the state, and, presumably, the restoration of elements of a competitive political environment. The new political course was focused on the political reintegration (as well as the utilisation of resources) of socially-advanced minorities (business and a new middle class) which the political regime during Putin's second term of office appeared to have alienated.

The Bipolar Structure of Governance

The main political effect of the electoral cycle of 2007–2008 was the reinforcement of the deconcentrated governance system, created by the anti-cyclic strategy in the process of managed rotation. A restrained transformation of the political regime was completed. The new configuration at the centre of political governance, which emerged after Dmitry Medvedev's ascension to power, finalised changes that had been started during the preparation for the elections of 2007–2008.

The most important developments happened at the "core" of political governance. Politically strong Putin left the restored institution of the presidency and became the prime minister. The prospect for a stronger role of government appeared. Simultaneously Putin headed the strongest party, having provided himself with institutional and political autonomy. At the core of political governance a second centre emerged. The symbiosis of the two centres displaced the monocentric governing structure. The system with the dominating centre was maintained, but the centre of the system became more complex: now it is bipolar. Deconcentration at the top generated a "disjointed" kind of domination, having separated the political component from the institutional one, and endowed two key institutions – the president and the prime minister as the chief executive – with new qualities. At first sight, the position of a strong prime minister, represented by a dominant political player deprived of his institutional properties (wide presidential powers) became the main innovation.

The strength of the prime minister is defined by his powers as head of the government, backed by wide electoral base, the support of the leading party in the State Duma, the political control over regional parliaments, and the retention of close links with the key figures of the governing group and its immediate administrative and political setting, including the ministers from "the power block" (defence, security and emergency agencies) and the CEO's of the largest state-owned companies.

But the presidency also changed: the new president retained all constitutional powers but was not the most powerful player; he did not have direct control over key political resources. For the first time the Russian presidency lost (at least for a while) its patronal attributes. (Now they can be executed only by mutual consent, as result of coordination between two key players.) The president became less powerful in political terms, but when we compare this to his previous "tsar"-like status in a monocentric regime, there are reasons to speak about the "normalisation" of the institution of the presidency. With the new political properties of the president and the prime minister in mind, the "disjointed" domination can be interpreted as the start of a transformation of the political regime. The effects which grasped both centres of governance are also important: deconcentration and depersonalisation of power became reality.

The promotion of two non-petersburgians, Igor Shuvalov and Sergei Sobjanin, to key positions close to the prime minister induced changes in the composition of the governing group. Taking into account the important role that the third non-petersburgian Vladislav Surkov played for a long time in the administration of the president, it is possible to speak about an increase in the proportion of other groups at the centre of the political governance system, which since 2004 had been considered as an exclusive domain of the petersburgians. At present only preliminary assessments could be made, but the extension of this trend opens the possibility of an evolution of the governing group along the route which may be called "from clan to class".[21]

Several changes in the structure of the ruling elite also became apparent. The most pronounced was the relative decrease in the status of the siloviki and the emergence of a new group of influential players - the judiciary and the chiefs of the law-enforcement agencies.[22] Public statements about the termination of the expansion of the state into the economy - together with a number of practical steps that were designed to confirm the value of the market system (for example, the reform of corporate governance in state-owned companies and the abolition of the privileges of state corporations) - opened

21 A. Zudin, "Novaya real'nost'. Osnovnyie napravleniya razvitiya politicheskoi situatzii v 2004–2008 godah" [New Reality. Main Trends in the Development of the Political Situation in 2004–2008]. Moskva: Tcentr politicheskih technologii, June 2004: http://www.politcom.ru/2004/prognoz12.php.

22 O. Kryshtanovskaya, "Bolshoi perehod. Uiti chtoby ostat'sya" [The Big Transition: To Leave, To Remain], Vedomosti, 23/04/2008, no. 74 (2096).

the prospect for enhanced status of private business elite. New directions in public policy (anticorruption policies and, probably, a new regional policy) created favourable conditions to the new round of political change – the reform of elites, their functional and institutional domains.

The main side effect of deconcentration and bipolar governance model was the appearance of institutional tensions. The joint governance of the president and the politically strong prime minister supported by the dominant party produced a political-institutional configuration that is not congruent with the super-presidential (patronal) system.

The Prospects of a Rupture with Monocentrism

The Future of the Bipolar Model

By consensus opinion, the immediate prospects of the political regime after rotation are determined by the stability of a bipolar organisation of governance. In general terms, the future of the new model, created by the managed rotation, can be described by two alternative scenarios: consolidation or disintegration of a bipolar regime.

The disintegration scenario is discussed most often. Its popularity could be explained by the fact that bipolar governance is rather unusual and also by the high conflict potential at the top, associated with joint governance.

The disintegration scenario may be provoked for three principal reasons. The first one: Medvedev's enhancement (gradual familiarisation with the "patronal president's potential") and Putin's attenuation. The second one: Medvedev's inability to obtain adequate political weight to balance strong Putin. The inability to be 'embedded' in an institutional model in which the president acts as the ultimate power-holder can lead to political destabilisation: in this case the new president will not be perceived as the key decision-maker by public opinion, elites and foreign leaders.

The disruption of coordination inside the joint governance at the top – caused by leader disagreements, team rivalry, or significant political asymmetries – could be the third reason for disintegration. In this case the regime breakdown potentially could be the most destructive since it implies polarisation and disintegration of the governing group with little prospect of reconstruction. On the contrary, the first and second versions of disintegration (actually, based on the reunion of institutional and political component of the domination) cannot generate 'fatal' consequences as they provide at least some sections of the opposite team the opportunity for to join the winner.

The consolidation scenario is discussed less often, but the devices created in the process of the managed rotation work to its advantage. Among them: mutual trust and psychological rapport (created by a long experience of teamwork), dense interlocks which interconnect the teams of the two leaders, a jointly elaborated political programme, cohesive effects generated by the presence of politically strong ex-president as the prime minister, resource dependence of the new president upon the old, the separation of control over institutional and political resources, the complementarity of the institutional and political centres of governance, which make it impossible to make decisions that ignore or infringe upon the rights of the other party (e.g., Putin's refusal to intervene in Medvedev's presidential prerogatives in combination with the recognition by Medvedev of Putin's political leadership), the successful "rooting" in the office and familiarisation with constitutional prerogatives by the new president, the organisation of a reliable system of coordination between two centres of governance.

In the case of disintegration, the new political properties of the system – political strength for the prime minister and normalisation of power for the president – would be lost. Institutional inertia would recur and (depending on how disintegration came about) the regime would return to one of the two already familiar political conditions – monocentric or "polycentric". But if the bipolar model remains stable (at least until the end of the new electoral cycle of 2011–2012), the new political qualities of the president and the prime minister can create conditions for the gradual dissolution of the monocentric system and for the growth of polycentric institutions.

The Possibility of a New Institutional Choice

The managed rotation of power created a critical junction: the cyclic dynamics of the patronal presidency were prevented, but the configuration of institutions which informed it was not changed. The effects of institutional inertia which the patronal presidency creates cannot permanently be neutralised by anti-cyclic strategies. If the institutions do not change, the same traditional pressures will reappear in the next electoral cycle.

Anti-cyclic strategies generate their own negative side effects: if the next federal elections are held in the same plebiscitarian style, this could could irreversibly modify the expectations and motives of voters. The campaign of 2007 strongly deviated from the institutional pattern: the Duma elections became a referendum of trust for the national leader, which was carried out by popular voting for a political party with which the national leader did not completely identify himself.[23] The movement away from the point of political equilibrium, fixed by the bipolar model of governance, is possible in two directions: the return to the logic of patronal presidency (along the two trajectories under the disintegration scenario) or the transformation of established institutions.

From the very beginning post-Soviet institutional transformation was driven by political domination and "imposed transitions". In this sense it is possible to consider the transformation of the political regime which began in 2000 either as the third imposed transition, following those of 1991 and 1993, or the third phase of the large imposed transition that began in 1991. The first imposed transition in 1991 was caused by a political catastrophe – an unsuccessful attempt to use 'force strategy' against reformers, followed by the disintegration of the conservative camp and the dissolution of the USSR. The reformers found themselves in the position of involuntary winners. The popular mandate was ambivalent, because it combined both the endorsement of "parade of sovereignties" (Russia included) as well as a desire to save the Soviet Union, expressed at the referendum of

23 A. Zudin, "Tsena voprosa" [The Question of Price], Moskovskiye novosti, no. 47 (30/11–06/12/2007).

March 1991. With a fragmented elite it was difficult to achieve public consensus.

The second imposed transition of 1993 was also triggered by the use of 'force strategies'. But this time there was a real winner (President Yeltsin). The popular mandate was also less inconsistent, so imposed transition was partly authorised by public opinion (the results of the March 1993 referendum are well known: "yes – yes – no – yes"[24]).

The third imposed transition began in 2000 and has not yet come to the end. The big difference is that the political polarisation of the elites on the threshold of transition (the electoral cycle of 1999–2000) was not accompanied by the use of 'force strategies', and the transition appears to be slow, gradual and a prolonged one.

In all cases, it was the governing group (not the counter-elite or political opposition) who initiated the change and took responsibility to shape the model of political development. (In the first case, however, political responsibility was practically imposed on the 'winner', and the task was not to 'shape', but to change the model completely.) Political, ideological and personal changes notwithstanding, it is possible to say that the successive governing groups kept at least minimal programmatic continuity. (In this respect it is also possible to speak about at least a punctuated line that connects Gorbachev and Putin: Putin's transformational agenda at the beginning of the 2000s can be considered as the continuation of not accomplished parts of Gorbachev's reform agenda at the end of the 1980s.)

Unlike the first two, the third imposed transition is to a much larger extent centred upon both the main trends of public opinion and established political practices. By the end of the 1990s, Russian elites had made substantial advances in becoming familiar with and subsequently modifying the imported political and economic institutions (e.g., insider-dominated privatisation, personalistic regional

24 The questions in the referendum ballot were put in the following sequence: 1) Do you have confidence in Boris Yeltsin, the president of Russia? 2) Do you approve social and economic policies of the Russian President and Russian Government since 1992? 3) Do you consider ahead-of-shedule presidential elections necessary? 4) Do you consider ahead-of-shedule elections of the People's Deputies of the Russian Federation necessary?

political regimes, authoritarian adaptation of elections etc.).[25] In turn, Russian society by this time expressed deep dissatisfaction with the established political system. Public hostility against imported institutions and new practices grew substantially and was manifested in the expansion of negative attitudes to the presidency as an institution, re-endorsement of the one-party system, renationalisation, and nostalgic feelings about the Soviet Union).

All the same the third transition retained some attributes of being "imposed": it was initiated from above by the governing group, accomplished with relatively high autonomy from elite and mass pressures, and involved the selective use of 'force strategies' (or a threat to use force). The third imposed transition began with substantial political caution: changes were implemented within an established constitutional framework, concessions were made to elites as well as to mass groups of voters, pluralism and political competition were restrained, but not abolished, and the strategic orientation to "modernity" was retained. (The authenticity of these statements is strongly challenged by the political opposition and many academic researchers.)

The presumable purpose of the third transition is to lay a foundation for a system of governance strong enough to cope with the regular personal alternation of power. If this assumption is correct, then the managed rotation of 2008 is only the first step. The ability of the dominant player to restrain himself from the temptations of the 'third term' basically makes the institutional reform possible. And if a two-poled model of governance is not a result of contingency, but rather a product of calculated choice, then we can expect a number of future political actions which could fix the new political qualities for the president and the prime minister in institutional terms. It is possible to specify two rounds of institution building on the escape route from the trap of the "patronal presidency" and "winner-take-all" situations.

25 Afanasyev M., "Klientelizm i rossiiskaya gosudarstvennost" [Clientelism and the Russian State], M.: MONF, 1997..

The "party cartel". This direction is closely related to the conversion of United Russia into the pillar of the new party system. Despite Vladimir Putin's participation in the Duma elections at the top of United Russia party list, the monopoly over the political market was not restored after the end of 2007–2008 electoral cycle. Fair Russia reaffirmed its location within in the exclusive "Kremlin niche". In 2008–2009 President Medvedev initiated a series of changes aimed at preserving restricted party pluralism and a moderately competitive political environment as well as introducing new procedural devices in order to increase executive democratic accountability. Presumably, the competitive environment (at least in a limited form) could help to reduce the effects associated with a hegemonic type of political dominance and stimulate the transformation of United Russia into in a 'real' political party (reinforcing holders of electoral and communicative skills within the party and squeezing out bureaucratic elements that are addicted to both their monopoly position and the usage of administrative resources).[26]

The cumulative effect of changes could create a new party system of the "cartel" type (initially organised vertically, but basically flexible both in internal proportions and the composition of participants) that is embedded in a consensus of norms and values. The domination will lose its institutional component and become mainly political (electoral). The political regime will finally cease to be monocentric but retain 'hybrid' characteristics, and the system of governance will be transformed to a presidential/parliamentary model. Other than classical hegemonic parties, the party cartel basically will be capable of solving the problem of uncertainty and the elites will regularly be confronted with the prospect of a rotation of power. At the same time the political cartel (with several participants instead of one) will reduce the dangers that are associated with political monopoly (especially 'the principal-agent' problem). But the implementation of this "party cartel" scenario is impossible without institutional reform.

26 Negative effects of the "hegemonic" parties are examined in: V. Gelman, "Perspektivy dominantnoi partii v Rossii" [The Prospects of the Dominant Party in Russia], Pro et Contra, no. 4, July–August 2006.

Reform of the electoral cycle. Favourable opportunities for the "party cartel" to emerge could be opened by the deconcentration of the electoral mechanism. In Henry Hale's concept, the patronal presidency itself creates a winner-take-all situation and aggravates the price of victory and of defeat to the extreme with a powerful impact on the motives and behaviour of the political players. But there is one more factor which produces high uncertainty and risk for the Russian elites: it is the dual electoral cycle which brings time limits of parliamentary and presidential elections in close proximity. The electoral mechanism becomes concentrated and focused on the presidential elections: parliamentary elections, as second-order elections, lose their independent value and appear to be subordinated to the logic of the presidential elections ("elite primaries", by Olga Shvetsova's definition).

In the 1990s the concentrated electoral mechanism enhanced the political strength of incumbent reformers. At the same time it reduced the power, autonomy and efficiency of other institutions. After a new round of political reforms from the early 2000s, hierarchical subordination of elections basically lost sense. The monopolisation of the State Duma by United Russia after the parliamentary elections of 2003 was in many respects a side effect of the super-mobilisation for the presidential elections of 2004. The retention of dual electoral cycle became counterproductive both for the presidency, overburdened with excessive power and responsibilities, as well as for impaired State Duma and political parties.

The idea of reforming the electoral cycle was first proposed by Sergei Mironov, the speaker of the Federation Council, who suggested increasing the President's term of office. (The proposal was publicly supported by Vladimir Putin in June 2007.) The immediate outcome of this proposal – in case it is implemented – is an enlarged interval between presidential and parliamentary elections. This purely procedural change contains strong political and institutional potential. Shifting presidential elections by a year (a term of five years) weakens the link to parliamentary elections. And if the time interval for presidential elections is increased to two years (a term of six years), the link with parliamentary elections would be broken completely. The disconnection of an electoral cycle opens up great opportunities. Parliamentary elections will be released from the grasp of presidential elections. The relative value of parliamentary elections will be increased. It would be possible to carry out two separate national electoral campaigns independently. The electoral formation of parliament and the presidency would be separated politically. The goals, motives, expectations and strategies of the players would be restructured and refocused on a particular kind of national electoral campaign. Disconnected electoral cycles would abolish the winner-take-all situation and enhance the prospect of normalisation of political institutions. The simple increase in term in office arranges an escape route from "patronal" (super-concentrated) presidential power.[27]

In case of success, the party cartel and the reform of an electoral cycle scenarios could consolidate the new qualities of the president and the prime minister, complete the separation of the institutional and political components of domination (which started in 2008), and dismantle the monocentric regime.

The Start-up of the New Transformation?

At the end of 2008 the first substantial amendments, introduced a month earlier by the new president Dmitri Medvedev, to the Russian Constitution of 1993 were adopted. According to the amendments, the terms of the president and the State Duma are extended to six and five years, respectively. Sergei Mironov's proposal was carried out, though in a revised and politically secure way. The deconcentration of the electoral process has already started, at least, in legal terms. In procedural terms the time interval between parliamentary and presidential elections was increased from several months to two years. The amendments will not affect the terms of current president and the State Duma and will take effect after the end of the electoral cycle of 2011–2012. It means that the next parliamentary campaign

27 A. Zudin. "'Yashik Pandory' ili vyhod iz instituzional'noi lovushki" ["Pandora's Box" or The Exit from the Institutional Trap?], Moskovskiye novosti, no. 23, 15/06/2007.

will start in 2016, and the next campaign for the president's office in 2018. From that time the political players will have to engage in two independent national electoral campaigns and the winner-take-all situation would begin to wither. The political effects of these constitutional amendments will enhance the institutional autonomy of the State Duma.

Another constitutional change would increase the strength of the Russian parliament: the federal constitutional law "On the Government of Russian Federation" was also amended at the end of December 2008. According to the amendments, the federal government is obliged to submit reports to the State Duma about its performance. In contrast to the amendments extending the terms of the president and the State Duma, the procedure of government accountability is effective immediately. The first report of the government before the State Duma took place in April 2009. Though it was not particularly impressive (the deputies did not have enough courage to question prime minister Vladimir Putin), it is fair to suggest that after the head of government is changed, the newly established accountability procedure will be transformed from a purely "technical" to a political one. This means that the political effects of all constitutional changes are suspended in time, but could be cumulative and strong enough to withdraw the Russian State from the logic of post-Soviet political cycle.

Galina Michaleva

Liberal Opposition Under Authoritarianism

Regime Changes and Institutional Conditions for the Opposition in Russia

The logic of development of a "fake democracy", i.e. an authoritarian state with the formal presence of democratic institutions is related to the deconsolidation of democratic institutions and their delegitimization in public opinion. The results of elections become predictable and are principally not changeable. In order to strengthen its position, the ruling group creates a system of shielding mechanisms and includes a propaganda machine This creates local and fractional repressions of those people who represent a real or potential (even weak) danger to the ruling system. To ensure the compliance of representatives of the political and business elite, ruling powers organize staged repressions and create external and internal enemies to justify such actions. Such actions are authorized by a state-controlled judicial authority. At the same time, blocking legitimate ways of regime change leads to intensified latent conflicts and an increased pressure on a society and stimulates the emergence of still very weak forms of resistance.

Democracy as a transitional solution[1] ceases to satisfy the dominant actor who strives to strengthen his position. Even democratic institutions established and guaranteed by the Constitution begin to fade away. Competition vanishes from the election process. "Delegate," "defective," limited, monitored, fake democracy is replaced with apparent authoritarian formations. Furthermore, applying Sartori's

1 Пршеворский А., Демократия и рынок: политические и экономические реформы в Восточной Европе и Латинской Америке. М., 1999. pp. 92–93.

criteria for the identification of authoritarian and totalitarian systems[2] in our analysis of the characteristics of a regime at this stage shows that the regime becomes more and more austere. There is no holistic totalitarian ideology as such. However, there have been attempts to develop such an ideology and its basic ideas, for example, "energy superpower" and "sovereign democracy". Potential and somehow real state control exists in different public areas. The number of these areas continues to increase and includes politics, business, culture, education, and public organizations. If repressions of a non-legal nature (force by terror, to use Sartori's term) are employed relatively seldom and selectively (Chechnya, Blagoveshchensk or show arrests and trials of businessmen, governors, mayors, as well as so-called "spy trials"), the threat of the use of such repressions (force by fear) is spread widely by the mass media. The subordination of social subsystems to the state increases.

We can talk about:
- total subservience of the most influential mass media sources, first of all television. At the same time, pressure on less important and local mass media increases. Attempts, partially successful, are made to control internet publications and blogs[3];
- the absence of independent jurisdiction;
- incorporating the Orthodox Church into the state system in exchange for some privileges;
- first steps in controlling education, such as the abandonment of election procedures of university rectors and the standardization of educational programs.

The next indicator is the isolation or exclusion of so-called "marginal" social groups. The illustration of this is the persecution of persons of "Caucasian Nationalities" and especially the anti-Georgian campaign of autumn 2006 and indirect fuelling by the

2 G. Sartori, "Totalitarimus: Modellmanie und Lernen aus Irrtümern", in E. Jesse (ed.) Totalitarismus im 20. Jahrhundert. Bonn, 1995. pp. 538–556.

3 R. Orttung, "Kremlin Systematically Scope of Russian Media", Russian Analytical Digest, Research Centre for East European Studies. Bremen, 9, 2006, pp. 2–6; D. Vinogradov, Russian Internet Remains an Island of Free Speech and Civil Society. Bremen, 2006. pp. 12–16.

state of nationalistic feelings and a policy of exclusion of sexual minorities.

The most obvious cases of power abuse are the anti-oligarch campaign, destruction of YUKOS, the imprisonment of Khodorkovsky, the selective persecutions of regional elites, and the ban or sanctioning of public demonstrations. Examples of the latter are the sanctioning of the 4th of November 2005 march of the Movement Against Illegal Immigration and other radical nationalistic groups in Moscow and the ban of an anti-fascist march of democratic parties and human rights organizations at the end of November 2005.

The growth of influence of the ruling party is obvious in spite of some disagreements on how to achieve the party's goal, that is, secure domination of the party entitled "United Russia."

The elements of boosting the key role of the "chieftain" are in place. This is exemplified not only by the wide public support of Putin but also in the conscious and unconscious construction of his "cult" (the construction of the image of a leader equal to the world leaders of G8; the fashion among officials to wear their watch on the right hand and to hang portraits of the leader on the walls of their offices, etc.).

Finally, state control of the economy is increasing by means of including authority officials in the upper echelon of management of the largest firms and the purchase of majority holdings of gas and oil company shares by the state.

By default, it is assumed that a wide consensus regarding regime-proclaimed values among the elites and the public exists. Such a "consensus by default" is seen by the ruling groups of elites as state-centralized, with the elements of nationalism and neo-imperialism, thus contradicting the spirit and letter of the current Constitution.

Prospects of the formation of new parties are being blocked and, as a consequence, the statements and representation of the interests of newly emerging groups in society as well.

The privileges of administrative parties and increasing restrictions for opposition parties cement the party system and stimulate the emergence of a radical non-systemic opposition which acts outside institutional frames. On the other hand, the eviction of all civil opposition parties to the periphery of the political process pushes them into collaborated actions of protest in election supervision and towards the forming of coalitions in legislative organs.[4]

Finally, the renunciation of the competitive election process and measures directed at strengthening the positions of the ruling coalition mean final renunciation of parties as democratic institutions which tie up authorities and society. Parties are gradually turning into an exclusive instrument for self-organization of the ruling elite group. The realization of the interests of the ruling party minimally depends on the will of the citizens.

The level of autonomy of systemic parties is minimal. At the same time, administrative parties and projects are connected with various, sometimes competing, factions of the ruling group. Beside one center of power, substructures emerge. These substructures are subordinate to the center and administrative parties depend on them. This results in competition among substructures. Such competition does not relate to pragmatic positions and political trends. Systemic public parties find themselves in a complicated situation of contradictions between pragmatic positions and suggested alternative trends. At the same time, they are subdued by the ruling group. The autonomy of non-systemic opposition forces is also illusionary. It shows in their regular activities – registration, appearances in media, receiving permission from the ruling group to conduct various actions.

The idiosyncratic feature of the political process is now its imitational character which leads to the virtualization and theatrical character of public politics.[5] The creative potential of the politicians and parties become a more important resource than the program, ideological positions or alternative political courses. It is not by chance that the popularity of Zhirinovsky is not decreasing. Deputies, according to journalists' opinion, "compete" in a level of absurdity of proposed bills[6] (from the requirement for women to obtain

4 Сергеев С. Политическая оппозиция в современной Российской Федерации: федеральный и региональный аспекты. Казань, 2004.

5 Ханукаева А. Партия – партийный лидер – общество: проблемы коммуникации в современной России// Власть, 10, 2006, pp. 20–24.

6 Рейтинг или смерть // Деловая газета Взгляд [Электронный ресурс]. Режим доступа: http://www.vz.ru/politics/2007/1/5/62758.html, 5/1/2007.

permission from their husband to have an abortion to going back to the Julian calendar). The sole goal of these is to appeal to the mass media.

Changes in electoral legislation and norms regulating the activities of the parties, as well as in the information politic of mass media under state control, are directed at guaranteeing the victory by the dominant party in elections on all levels, and at minimizing the role of other parties and, first and foremost, of the parties of the opposition.[7]

The consistent and purposeful reduction of the zones of freedom and the creation of limitations of democratic institutions are accompanied by state rhetoric and, as it seems, contribute to the strengthening positions of the ruling party. The growth of prices in energy supply resources has created exceptionally favorable conditions for any moves inside the country as well as abroad.

However, the disparity of positions on the one hand and certain self-restraints of the authorities on the other do not allow the regime to be consistent and to begin the ultimate liquidation of the actors which are a potential or realistic threat to the monopolistic position of the dominating actor. These competing forces include representatives of business groups, regional elites, opposition parties and movements, even though weakened in their state.

Contained within the realities of a consistent strengthening of the authoritarian-bureaucratic regime in Russia, there are several factors which conditioned the trajectory of development of the party system and the strategy and tactics of the opposition forces.

The rules of the game have changed and this has affected both formal and informal institutions. A body of new legislative acts, including changes in the party law and electoral legislation, have completely altered the electoral formula and set up new conditions for the parties. The Kremlin administration and a guaranteed parliament voting majority consistently acted in accordance with the goal of cleansing the political field. They created most favorable conditions for the party designated to become dominant in the political

7 Гельман В. Трансформация российской партийной системы: монополизация политического рынка// Российские элиты в условиях консолидации власти. Пермь, 2006. pp. 27–47.

system[8] and most unfavorable conditions for all other parties. In other words, a party system under total control was created.

In December 2006, a Duma majority passed amendments to the party law. These amendments made the difficult conditions of the parties' existence even more so and made it practically impossible to create any new parties which would not be subordinate to the Kremlin.[9]

Requirements for the membership quota increased (Article 1): at the moment, a party should have no less than 50 thousand members. Some party organizations should have no less than 500 and others no less than 250 members. Article 2 indicated deadlines to sort out compulsory numbers and branching in accordance with the requirements of the law of 1 January 2006, i.e. parties had to meet the new requirements a long time before the beginning of the election campaign. Parties who did not satisfy the requirements of the law on parties were ordered to self-liquidate and to become public organizations. Otherwise, they would be liquidated by a court order. Passing of this legislation meant the last hour for small parties, which practically lost all rights to exist as political subjects.

The strict norms of the law on parties were complemented by changes in the electoral legislation. The following became key norms, which practically altered the electoral formula:

1) The introduction of a 7% blocking-off barrier on a federal level and, following this, the 2006 change of regional electoral legislation which raised barriers in regions (most often 5% but in some regions, for example, Primorje 3%) up to 7%.
2) The introduction of a proportional electoral system on the federal level. The federal list should not include more than 3 names and the list itself should be split into 100–150 regional lists.
3) A ban on the participation of public organizations in electoral units.

8 Сурков потребовал от "Единой России" обеспечить доминирование партии на 15 лет// Lenta.ru [Электронный ресурс]. Режим доступа: http://lenta.ru/news/2006/02/27/dominate/, свободный.

9 Ст. 5272// СЗРФ, 52, 2004.

4) A "tacit" option for the parties represented in Duma to nominate candidates for elections on all levels. Complexities for other, non-Parliament, parties to nominate their candidates.
5) Raising of the election tax from 37.5 to 60 million roubles and subsequent corresponding changes in regional electoral legislations.
6) Lowering the allowed level of "defective signatures" from 25% to 10% and imposing constraints upon organizers of activities to collect signatures to nominate a party make such activities virtually impossible. The practice of depositing already obtained signatures while continuing to collect others in order to counteract the negative influence of defective signatures was also banned.
7) The introduction of an imperative mandate and the loss of deputy status when leaving the faction.
8) A ban on blocs and then a ban on participation of members of one party on the lists of another party.
9) Preservation of the right of "parovoz", i.e. a governor heading federal and regional lists can pass his mandate to a candidate following him on the list (de-personification).
10) The right of the party that wins regional elections to nominate the candidate for the position of governor.
11) Amendments to the law "On Counteracting Extremist Activities" which expand the definition of extremism and, practically, allows officials to label any opposition and criticism towards the ruling power as an extremist activity.
12) Lifting restrictions from pre-term voting.
13) Cancellation of voter's turnout margin.
14) A ban on the criticism of other parties and competitors in the election process.

These amendments, which seemed to be separate measures, in sum have led to a constriction in the range of activities of those parties that are not related to the ruling power.[10] A change of the ruling power at the elections is thus very unlikely.

10 Макаренко Б. Российская избирательная система в контексте эволюции политического режима// Политком.РУ: информационный сайт политических комментариев [Электронный ресурс]. Режим доступа: http://www.politcom.ru/article.php?id=2812, свободный.

Another consequence of these norms, which made elections campaigns easy for administration parties and difficult for opposition parties, is the practical loss of enormous numbers of citizens to both active and passive electoral legislation. This is not Constitutional.[11] It is noteworthy that an institutional choice allowing the minimization of political competition appeared to be insufficient. The logic of development of the authoritarian regime forces power groups to guarantee their dominant position through legislation. Potential competitors have less zones of freedoms and less options. A combination of the president's popularity among citizens and his acceptance by the international community, high oil prices and thermidorian feelings among the elite (to instil order, to strengthen the chain of command, to restrain those who had gotten too high in the previous decade) could not last forever. This is why, any "struggle of sections", or sections that were too autonomous, were perceived as a threat to Putin's consensus of elites.[12]

Elections, which have already lost any feature of competitiveness, are gradually losing the option of an alternative. Parties are presented with some kind of "property qualification" or "administrative qualification."

By the end of 2004 and beginning of 2005, the registration service of the Ministry of Justice began full audits of parties in regions. These audits did not only check the numbers of members through documents and through selective telephone calls and visits of the enlisted members, but also checked meeting minutes, including local meetings, and their conformity to legislation and the party charter. The demands placed on the parties could only be satisfied by state institutions with a permanent active bureaucratic apparatus.

The inevitable consequence for the parties: beside United Russia and the parties who were consigned with a special function (for

11 Вишневский Б. Право как воля господствующего класса// Политический журнал, 23, 2006, pp. 4–7.

12 Макаренко Б. Политические партии постпутинской эпохи// Политком. РУ: информационный сайт политических комментариев [Электронный ресурс]. Режим доступа: http://www.politcom.ru, 15/8/2006.

example, spoiler) by authorities, only those parties which managed to provide the necessary documentation and firm membership base survived. That was true in those cases when there was no command to attack because no public organization was able to meet all the requirements. Members of the opposition parties must have courage and firm convictions not to give up their party membership under the current circumstances.

Meanwhile, those who did not have a chance to survive as an independent player fell into a worsening position. For those who stayed in the field, the chances increased, and any incentives to join in with the weaker players disappeared.

Furthermore, the new electoral formula allows the merger without complex amalgamation of two or three parties into one with the inevitable program compromises and mutual concessions and lowering the status of the management. Those who lost the status of a party and consequently could not participate in elections, had to agree to conditions dictated by those who remained in the political field or to "sell short" and proclaim regime and election non legitimate.

At the end of 2006, the Central Information Office officially announced that 19 parties underwent audit and 16 parties were found not to satisfy the requirements of the legislature.[13] All parties who were refused registration insisted that the audit procedure violated federal legislation and were very critical of the process. "This was not an audit but an attempt to fulfill the state order to liquidate unwanted parties," said Viktor Cherepkov, the leader of the "Freedom and Democracy Party".[14]

Those who did not pass the state registration service audit chose several different strategies. The independent deputy of the state duma Vladimir Ryzhkov, who had hopes of resurrecting the Republican Party and planned to head democratic opposition, contested the audit results in court. After he lost his case in the Supreme Court, Ryzhkov appealed to the Strasbourg Court of Human Rights.

13 В России только 19 партий допустят до выборов// Подробности.ua [Электронный ресурс]. Режим доступа: http://www.podrobnosti.ua/power/elections/2006/10/26/361333.html, свободный.

14 Магомедова М. Принудительная ликвидация // Новые известия. 2007. 12 янв.

Parties and Public Organizations on the Liberal Side of the Political Specter

The parties "YABLOKO" and The Union of Right Forces, who lost Parliament status in December 2003, became targets of immense criticism of journalists and politicians who were in opposition to them. Political scientists and political technologists evaluated their chances for the future as minimal.

СПС [The Union of Right Forces] announced a management change. The leaders of the party resigned and I. Khamakada left the party. СПС, still possessing huge resources, did not have a leader for a year and a half until the party discovered a promising backwoodsman from Perm, Nikita Belykh. The party gained a few victories during the first regional elections; however, the number of successes was decreasing. The main problem of СПС, which supported the president up to the state Duma electoral campaign of 2007, was its political identity. Domination by United Russia, which still has a liberal wing, put into question the necessity of having another player on the same field. The attempts of Nikita Belykh and his supporters to express opposition views placed them in close quarters with "YABLOKO" and reduced their chances for continuous funding. A. Chubajs, who was directly subordinate to the Kremlin, could not finance the party from his business projects. СПС leaders openly stated that they did not have any chances to get over the 7% barrier in 2007. After the Moscow elections, a few representatives of СПС conducted separate negotiations with "YABLOKO" leaders and were ready to leave their party in exchange to obtaining significant positions in "YABLOKO". During informal agreements of leadership representatives in 2006, the parties tried to "separate" regions. Until the change of legislation and the introduction of the ban on participation in the lists of other parties, СПС members participated in some regional elections on the YABLOKO ticket. СПС was on the verge of a split in ranks. However, after the СПС list received 16.5% votes in the elections of Perm Region in December, the СПС Congress of 16 December 2006 announced that the party woud stop negotiations with "YABLOKO" and was planning to participate on its own in all regional elections. A. Chubajs and, following him, СПС leadership dependent on financial and administrative support, chose to keep a political structure that was loyal to the Kremlin and to dissociate themselves from the opposition. The subsequent regional elections were conducted and the agitators were paid commissions for gained results. The major theme of the elections was the increase of retirement pensions. СПС began to work on an alien ideological field and this angered communists and Fair Russia.

"YABLOKO", in spite of vigorous internal party discussions, preserved its position and its leadership. The party managed to survive practically without financial resources up until the end of 2005 due to the change of the algorithm of interaction between party leadership and regional organizations, changes in pragmatic statements, activization of relations with civil organizations and protest actions. The absence of resources, however, did not allow the party to gain significant victories in regional elections. By the middle of 2006, the party managed to acquire legal status and run a unification congress.

During Vladimir Putin's second presidential term, the time of curtailing of democratic institutions and changing of the vector of transformation, a situation emerges which is similar to the period of liberalization, that is mobilization and politicization of Civil organizations. The difference is in the correlation with the vector of political development: during the period of democratization, non formal organizations were fighting for their freedom and regime change; now, legal public organizations are forced to defend freedoms and rights obtained earlier and defend their gradual loss of autonomy against the state. Ruling powers, who were retreating then, even though not without resistance, have in their possession at the moment different institutions, selective repressions and imitation (fake) institutions.

Public organizations remained on the periphery of the ruling powers' attention during the entire period that followed the regime change and election of democratic institutions as transitional solution. Public organizations were used as a supplementary resource or an instrument to attract of western donors.[15]

15 Граждане и власть: проблемы и подходы. М., 2001; Шмидт Д. Какое гражданское общество существует в России, Pro et Contra, 1, 2006, pp. 6–24.

During Putin's first presidential term, the Civil Forum, an attempt to regularize public organizations and to use them as supplementary resource to legitimize the regime, failed. Civil and especially human rights organizations were financially independent of the ruling powers; they used western donors and were not eager to line up in the institutional chain of command. On the other hand, the ruling powers did not have time for these organizations then. More important problems had to be solved in order to consolidate power: re-centralization, the building of a manageable parliament and party system, the subordination of business groups, etc. During Putin's second term, these tasks were already fulfilled. The colour revolutions in neighbouring contries demonstrated that HKO (non-profit organizations) can become the nucleus of organized public protests which might lead to the loss of power by ruling elites. This is why the politics of segregation toward HKO came into being. It should be noted here that the elements of the following type of interaction – creation of common chambers under governors, creating obstacles for activities of human rights organizations, show trials – also took place earlier.[16] At the moment, the full-scale attack has begun.

First of all, the law on public organizations was ratified. This law considerably complicated conditions of organizations' existence and put restrictions on receiving funds from western donors, practically the sole source of funds. Critics of this law inside the country and abroad likewise noted that the ratified law contradicts the current constitution of Russia as well as international standards of people's rights to create public organizations and express opinion. The law considerably limits the autonomy of HKO and puts them under state control.[17]

Second, the Public Chamber was created, which received the exclusive right to act on behalf of civil society.[18] The Public Chamber is designed to play the role of a "proper" civil society and, beside

16 Петров Н. Общественная палата: для власти или для общества?// Pro et Contra, 1, 2006, pp. 42–44.

17 D. Taylor, "Law Enforcement an Civil Society in Russia", Europe-Asia Studies. 58(2) pp. 193–213.

18 E. Fein, Potjomkinisches Parlament und Papiertiger: Russland Analysen. Bremen, 2006, 87, pp. 2–4.

its "multifunctional" utilization, it can be used as a well-controlled stage for public politics and a supplementary mechanism for putting pressure on officials.[19]

Third, the threat to lose their positions forces western sponsors such as George Soros to act with caution or to give up their work in Russia altogether.

Fourth, a campaign to discredit most influential human rights organizations has been launched. In this campaign, familiar Soviet tricks are used - accusation of human rights organizations of western support and espionage activities.

The nature of activities of human rights and ecological organizations is such that they are supposed to influence the government on behalf of the public. These organizations now find themselves in a difficult position. Their activities, which were not considered political before, are now treated as such, and the organizations themselves are treated by authorities as political opponents.[20] The worsening of conditions for human rights organizations is forcing them to change their attitude toward liberal parties.

Back in 2003, "Memorial", the Moscow Helsinki group, "Golos" [i.e. Voice] and others rejected institutional collaboration with liberal parties because of the fear that they would be "used", and preferred only personal, even though close, contacts. On the other hand, human right activists worked together with authorities and participated in various councils on human rights under the auspices of the president, Duma, and they even participated actively in the Civil Forum, the first Kremlin's attempt to create "chain of commands of the Civil society" subordinate to regime.

Currently, the ruling authorities are practically "pushing" these organizations toward political activity. But by their nature and functions, these organizations are not capable of engaging in political activities. This is why they require political partners which have vertical structure and resources of various sorts.

Several different directions in cooperation between public organizations and opposition parties existed:

19 Н. Петров. Указ. соч. pp. 51–53.
20 Правозащитное движение сегодня: проблемы и перспективы. М., 2005.

1) The attempts of human rights organizations to found their own parties ("United People's Party of Soldiers' Mothers," "Green Russia," Human Right Activists (initiative of L. Ponomaryov) failed. These attempts stopped after thickening of the legislation (see below). Most of the organizations which planned to engage in their own political activity, have chosen some existing political party as "their party."

2) The creation of a new type of broad conglomerations which would include not only human rights organizations but also parties, political organizations and prominent politicians. Civil Congress was created in December 2004 on the initiative of a non-formal association "Za pravo bybora" [for the right of choice] (which included a number of influential НКО, e.g. МХГ and the association "Golos"), founded as a result of joint work on claims for judicial inquiry to challenge results of 2003 elections, and with direct participation of "YABLOKO". Representatives of all influential human rights organizations from various regions and prominent politicians of the democratic wing took part in the work of this Congress. a permanent operational network was established. After the second Congress and more than a year of joint work on various projects, representatives of human rights organizations regarded as rather natural their work in structures which included parties. СПС, whose representatives participated before in the Action Committee of ВГК [All-Russian Civil Congress], proclaimed itself a member of the Committee after Moscow elections. A number of human rights organizations participating in the work of the Action Committee increased. But positional disagreements among leaders and organizations, which are reminiscent of the times of split of umbrella movements, led to the alteration of the ВГК structure and its shift to the system of individual membership. At the same time, several politicians who do not have parties for various reasons (G. Kasparov and the United Civil Front, M. Kas'yanov and his National-Democractic Union, E. Limonov and his НБП) and representatives of human rights organizations set up the "left-right" coalition "Other Russia" in the middle of 2006. This coalition is a pretender to the status of major opposition force to the Kremlin. One year later, after a series of non sanctioned manifestations ("March of Dissidents"), the coalition has begun to fall apart. Kas'yanov and Ryzhkov and a few human right organizations have left.

3) The attempts of HKO representatives to act as mediators at the "unification of democratic opposition" under "Committee 2008" and subsequently, "Civil Congress."

4) The parties' search for allies among HKO and utilization of their networks to organize large scale public events (marches in defense of freedom of speech, anti-fascist marches, meetings, piquet) in Moscow and regions. КПРФ and "YABLOKO" declare that such actions are their strategic goal. СПС began activities of this sort only in 2006 and sponsors anti-fascist and human rights marches. Since restrictions have been imposed on public manifestations, all political forces are resorting more and more often to non sanctioned actions. The response of the ruling authorities was to resort to force and to employ pro-Kremlin youth associations in public street activities.

The inconsistency and non transparency of these processes remind of public movements during the period of liberalization. As in those days, the nature of interaction of organizations now is defined by the positions of leaders and by non formal character of relations. As in those days, conflicts and splits are inevitable and caused by the differing interests of individuals and the absence of institutional regulations. The sequence of events at the Civil Congress and around it in 2006–2007 is very illustrative of the above. This organization does not have a well-defined structure or mechanism for passing resolutions and following them up. This is why many parties, politicians and pseudo-politicians, especially those who failed to create their own political parties, envisioned Congress as a stage for establishing their political platform. The co-chairs of ВГК, G. Kasparov and G. Satarov, as well as M. Kas'yanov (the latter did not participate in ВГК's work per se) looked at ВГК precisely from these positions. The congregation "Other Russia", which was convened by "non-systemic" politicians from liberal to radical left, was meant to become a political constituent of ВГК. This development of events did not appeal to the systemic parties that participated in the Congress – "YABLOKO" and СПС. Serious conflict, which led to a no confidence vote for the co-chairs, resulted in the decision of Congress (after its third annual gathering) to accept the principle of individual membership. The task which these parties pursued, the depolitization of ВГК and delegitimation of the political ambitions of the above mentioned politicians, was fulfilled.[21] Both mentioned parties established their position in relation to "Other Russia" in 2007. The position is institutional non-participation in the Congress. Individual members, however, would not be banned from participation in "marches of dissidents" or making speeches at the Congress. This position became the reason for the accusations against both parties by "Other Russia" of their "concert with the Kremlin."

There were several positions regarding the future of democratic forces among democratically oriented politicians and organizations. The positions changed alongside changes in legislation and failure of new political projects.[22]

1) There is urgent need to create a new Democratic Party because "old democrats" have lost their influence (V. Ryzhkov, I. Khamakada, V. Lysenko and others). However, attempts to create new parties or "blow life" into old ones have failed. The proposed projects "Our Choice" of I. Khamakada, "Green Russia" of A. Yablokov, and even the revival of the parties of the first generation - the Republican Party of V. Ryzhkov and V. Lysenko and the Democratic Party of M. Kas'yanov - were not realized.

It appeared that the number of democratically oriented citizens in the regions is very small and those who are willing to participate in political activities are already members of parties or human rights organizations. The only way to create a new party is to recruit members of "YABLOKO" or СПС, who did not see any good reasons to leave their parties and create personal resource for second-rank politicians. Audits by registration bodies reduced to null efforts to increase numbers of party members. I. Khamakada rejected her project in its initial stage and joined Kas'yanov. V. Ryzhkov managed to attract V. Mel'nikova and her supporters from the "Committee of Soldiers' Mothers"; however, it was very difficult to build a party

21 Тирмастэ М. Партийные демократы разругались с беспартийными// Коммерсант. 2006. 12 дек. Позиции, связанные со стремлением политизировать ВГК отчетливо прослеживаются в ряде публикаций этого периода в сетевом издании: Ежедневный журнал [Электронный ресурс]. Режим доступа: www.ej.ru, свободный.

22 Россия Путина: руины и ростки оппозиции. М., 2005.

fast under the new strict regulations. The registration Committee brought a case of РПР liquidation before the Supreme Court.

Kas'yanov, who appeared to the Kremlin as the most dangerous person because he expressed his willingness to run for the president, was neutralized in a different way. ДПР, which was de-facto a semi-virtual structure and was offered for sale to prospective sponsors, was simply purchased.[23] Very pragmatic members of the party leadership, who were waiting for a sponsor but did not want to risk inevitable conflict with the ruling powers, made their choice. The same scenario as in the case of КПРФ split was used: an alternative congress. This time the plan worked out better. A. Bogdanov, who used to work in the Central Executive Committee of EP in 2002–2003, was elected a Chairman. ДПР became a spoiler and Bogdanov started sending letters to the leaders of regional organizations of "YABLOKO" and СПС where he called upon them to unify on the basis of ДПР and to abandon old leaders. Leaflets of similar contents followed but all this did not produce any results. Kas'yanov announced that he was forming a new movement – the People's Democratic Union – which would be able, in case of need, to become the basis for the creation of a new party. In 2006, Kas'yanov tried unsuccessfully to be nominated as a presidential candidate by the united right-left opposition. Kas'yanov played an active role in the project "Other Russia" up until his departure from the united right-left opposition in summer 2007, when it became clear that the number of prospective candidates was constantly increasing.

2) It is necessary to unite all democratic parties, first and foremost "YABLOKO" and СПС (G. Satarov, G. Kasparov, a few members of the leadership of СПС including B. Nadezhdin and N. Belykh).

The main goal for most observers and participants of the political process was "unification of democrats" and renunciation of mutual demands. The leadership of "YABLOKO" put programmatic principles and economic development of the country to the front while agreeing with the necessity of creation of a United Democratic Party.

23 А. Макаркин. Съезды демократов и республиканцев// Политком.РУ: информационный сайт политических комментариев [Электронный ресурс]. Режим доступа: http://www.politcom.ru/article.php?id=239, свободный.

Nadezhdin's suggestion to select a party which would serve as a matrix for the unification of democrats has been pronounced many times and best represents the position of СПС. G. Kasparov established "Committee 2008" in 2004 with the purpose of realization of such suggestion. The committee included well-known journalists, public figures, human right activists. Representatives of СПС (Nemtsov, Nadezhdin) and "YABLOKO" (Ivanenko, Mitrokhin), as well as I. Khamakada, V. Ryzhkov, G. Satarov, also participated in the work of this committee. The work of the committee did not yield any satisfactory results due to different visions of tasks and claims of second-tier politicians to absolute leadership.

The second attempt of this nature was the initiative of G. Satarov and G. Kasparov, the co-chairs of the All-Russian Civil Congress. At the second Congress, they presented an "ultimatum" of urgent unification of "YABLOKO" and СПС with everybody else. However, neither "YABLOKO" nor СПС needed such middlemen after elections to Moscow Duma.

The conference "Other Russia", which took place on the eve of G8 meeting in summer 2006, divided politicians into two groups on the basis of their adherence to certain democratic values: non systemic opposition – those who could not meet normative requirements for participation in elections and systemic opposition – those who had such chances or hoped to get them after their registration with the Ministry of Justice. "YABLOKO" and СПС distanced themselves from this enterprise. Kas'yanov, Khamakada and Ryzhkov announced the creation of a "permanent committee" at the conference. The participation of left radicals Anpilov and Limonov in this conference marginalized this side of political spectre and diminished chances for unification.

The above mentioned third All-Russian Civil Congress weakened the project even more. Kas'yanov and Kasparov practically remained in isolation and were shunned by political forces with large resources and some public organizations, especially human rights organizations, who considered collaboration with НБП and РКРП impermissible.

3) It is necessary to strengthen party's organizational structure and to attract active politicians and organizations to the party.

РДП "YABLOKO" declared and consistently implemented this position at the XIIth Congress in 2004. The majority of the party's leadership and regional organizations, unlike journalists and public figures, did not consider amalgamation with СПС the only possibility. They even thought that such an amalgamation might bring more harm than dividends. The strategy of "matrix" was not acceptable to a party with more than a decade-long history. "YABLOKO" consciously and consistently worked on the establishment of a unified party on its own basis. Interaction with network organizations within the framework of the Civil Congress, in which the party played the key role, and the structure of the "YABLOKO"-United Democrats" list laid out firm foundations for talks with "Green Russia", the Moscow Organization of Soldiers' Mothers, and human rights organizations. An internal split and uncertainty of СПС prospects led to the situation when СПС refused to participate in Moscow elections and СПС members were listed on the "YABLOKO" ticket. 11.1% votes given to СПС candidates allowed them to create their own faction (alongside ЕР and communists) in the Moscow city Duma and even though this victory did not give them political power in the Duma dependent on Luzhkov, it did have symbolic character.

The possibility to form factions and the assurance of procurement of important positions in the party, alongside thickening of electoral legislation and the law on parties, played the decisive role in that some members of "Green Russia" and Soldiers' Mothers expressed their willingness to join "YABLOKO." Factions started to emerge on federal and regional levels. This initiative gave impetus for politicized human right activists to apply to "YABLOKO." Youth and gender movements which did not have formal status inside the party, received formal status. The existence of five factions and some alterations of the party program, mostly, enlargement of ecology section of the program, allowed the party to announce the transformation of the party into united РОДП "YABLOKO." РОДП "YABLOKO" has become the most significant and, probably, the sole, player in this section of political specter. However, the party's further existence was threatened by institutional conditions under which the party functions, as well as by the complex position of the party's opposition to the system. The party's position of civil systemic opposition appears quite complicated for both voters and party members. The party faced the following dilemma: to continue to play according to the compulsive rules, to risk loss of face and to become one of pro-Kremlin pseudo-democrats or to radicalize itself and to lose systemic status.

The Growth of Opposition Feelings

The results of the State Duma elections December 2007 significantly damaged the position of the democratic opposition. No democratic parties overcame the 7% barrier. But in spite of the growth of state revenue, consistent implementation of policies that infringe upon the rights of citizens led in this period to huge street protests which have now acquired a methodical character. The most burning issues are price rises on housing and communal services, prejudices to the rights of certain groups of citizens (dormitory tenants, tenant builders, tenants protesting against infill constructions or green land clearances and so on). Some public events are organized, some are spontaneous. The major organizer of these events is usually КПРФ. Liberal parties and, above all, "YABLOKO", starting with 2004, consistently increase their presence in street public protests and later move from mere participation to their organization. The Union of the Right Forces [СПС] has also started to join public events, including non-sanctioned events.

The other option for democratic opposition is dealt with infighting caused by the transition of power from the present president to his successor. This infighting reduces the unity of political regime and therefore may create additional options for oppositional activity.

Spontaneous events which are reminiscent of the protests of the end of the 1980s are becoming citizens' reaction to local and national events. Infill construction in a certain house, loss of bank deposits in certain banks, the ban on the import of cars with the steering wheel on the right side, certain actions of regional and local authorities mobilize active citizens. The scale of the protests and the level of mobilization are not comparable with the end of the

1980s. However, they are becoming noticeable. The ruling powers continue to react according to the old scheme: restricting citizens' right to protest, changing legislation, resorting to administrative measures and even to violence. They begin to use methods tested in Chechnya: recourse to "special events" and filtration stations. One such case is the reaction of Bashkotorstan police to public street protests in Blagoveshchensk: 2.5% of the inhabitants of the area, or more than 1000 people, were taken to these stations and beaten up.[24] This forces even systemic opposition to recourse to non-sanctioned events every so often.

New forms of mobilization and new forms of actions are appearing. Participants can get together with the help of blogs. They use short spectator events (flash-mob) which are intended to attract journalists rather than passers-by. The number of youth and politicized movements participating in such actions increase.[25]

The ruling powers are also using new forms of reaction to protests. They have created pro-Kremlin youth movements ("Marching Together," "Nashi" [Our People], "Young Guard"). Regional authorities in regions where governors belong to the top echelon of EP have created similar regional groups (for example, the "Locals" in region of Moscow). The major function of these organizations is to participate in public events in support of ruling rowers.

The total annihilation of democratic institutions has not yet occurred. There is still a possibility of choice. The final renunciation of democracy as being a transitional solution, the creation of a chain of command with dominant party and subordinate parties – satellites and imitations of opposition – with no options for election results (transfer of power from the president to his successor or staying in presidency for the third term) will turn parties into an element of the authoritarian regime. In such a case, the system, due to its endogenous instability, will have to go back to the starting point of transformation.

24 Luchterhandt O. Jagdszenen in Russlands Provinz: Der Fall „Blagovescensk" in der Republik Baschkortostan// WGO, Monatsheft für Osteuropäisches Recht, 4, 2005, pp. 254–266.

25 Siegert J. Politische Jugendorganisationen und Jugendbewegungen in Russland: Russland Analysen. Bremen, 83, 2005, pp. 2–5.

Afterword

In the 3 years since these articles were written, the tendencies that were stated and analyzed by the authors progressed, being strengthened by exterior and internal factors.

The key events that determined the country's vector of development were:

- Solving the problem of the "successor" and the gradual transition towards tandem management while preserving the role of the former president, who became the prime minister.
- The Five Day War with Georgia and the following recognition of Abkhazia and South Ossetia as independent states.
- The financial-economic crisis.

Russia's modern political system can already be classified not as transitional, but as an established model with its own logic of internal development. The system's specific characteristics can be seen more vividly when Russia is compared with post-Soviet countries. During Soviet times, the more independent our government became of society, the more closed and secretive it became, but under Putin this trend reached proportions greater than that of early Soviet or post-Soviet times, comparable only to that of Stalin. Later, the mystery shrouded in Putin's "successor" operation gradually transformed into the mystery around the current "dual-rule". Putin's "successor" operation and its modern continuation have a lot of elements which are easier explained by a lack of exact plans and undefined goals than by Machiavellian cleverness and secret dealings. The lack of this plan, covered up with illusions of a secret plan, can be explained not only by Putin's characteristics but also by the situation in which he is in. In my opinion, this is related to stark contradictions, the "system of values" of the former president and the people surrounding him, which is a collection of contradicting views that cancel each other out. The unfamiliarity of the current situation and the complete unpredictability of our political future, the final result of which is chaos; the mixing up of our mass consciousness with the society's consciousness; the inability to live in democracy and the inability to abandon it at the same time: all these are due to the fact that there are no alternative models. Everything is frozen in a hung up state waiting for the administration's decision.

The state of contemporary Russian society, as the last study of the Levada-Center decisively showed, is specific in its apathy, impoliticness, orientation towards consumption, lacks solidarity, the perception of everything outside of one's own personal life, including that war, is like a TV movie or football match. The majority of the people, including those who are listed in "socially advanced" groups (high income, possibilities of career advancement) have, as a result of massive propaganda of opinions and worldviews - which were placed upon incompletely developed ideas of norms and values in post-Soviet society –, a "schizophrenic perception".

From one side there is deep distrust towards all political institutions and organizations, with the exceptions of Putin and Medvedev, as subjects of politics. People try to solve their problems by themselves (relaying on small networks of contacts - relatives or patron client relations) and not trusting the government, opposition parties or NGO's, and especially not the courts of law in these matters.

On the other side a perception of a paternalistic government that is required to take care of these exists, as well as state-centric perceptions concerning a person's place in the country and the world. Out of this comes patriotic delight, nationalism, xenophobic orientations, and the search for exterior enemies that relies on the post-imperial syndrome and the old world view where the West and NATO are the main enemies. Propaganda is received well on the basis of anti-Western and anti-caucuses sentiment. Conceptions of a renaissance of "national grandeur" work as a classic compensative mechanism. People undergo (illusionary) "delight from their own unity", when they see that the majority shares their happiness over football victories, victory over Georgia or over sharp commentary at the UN. Mass consciousness is fragmented, innovations coincide with tradition, while at the same time both are transformed, giving them a hybrid form. A significant part of the processes of mass consciousness should be classified with regard for their "dual" status, specifically the presence of both tradition and innovation in different forms. The elevated physiological meaning of attributes of tradition such as "the state", "border", "territory", and "the past" can be viewed as manifestation of neo-traditionalism,[1] and as a witness of the continuing modernization. The traditionalization of "general" political concepts is accompanied by cognitive and value acquisitions. Mass consciousness is more differentiated in a certain part (the character of the West) and more realistic in others (critical views of the role and "quality" of democratic institutions and a more realistic assessment of relations with the rest of the world). At the same time Russia's rulers are keen to use the traditional world view and values.

Citizens that do not have a clear system of values, orientations, perceptions of the past and present, and of the world's countries are left one on one with the propaganda machine. At the same time the government tries to use the situation to construct mechanisms of mobilization, which, in effect, distract people from the problems of society and the country of which they are a part of. However, these effects are short-term. The society's disposition can change quickly, especially against the background of the financial crisis. While these effects will be felt by the mobile (relatively) high-paid groups first of all, the lower income groups have already been hit by price increases on food products and unemployment. The government and the majority of the people are not ready to live in accordance with democratic values and rules even though they necessity of these is declared every so often. Disrespect for rules, freedom, and human rights are widespread characteristics. Elections have turned into a political farce and have practically become appointments. The courts are dependent on government and corrupt. Mass media have become a tool of propaganda. Citizens stopped believing not only government institutions, but also political parties, civil societies, trade unions, and even their neighbours. A deficit of trust and solidarity are obstacles to uniting for collective action and opposing the government. To say even more, the overwhelming majority of the

1 Гудков Л. Русский неотрадиционализм // Экономические и социальные перемены: мониторинг общественного мнения, 2, 1997, pp. 25–33; его же. Русский неотрадиционализм и сопротивление переменам. – "Отечественные записки", 3, 2002, pp. 87–102.

people are still ready to forfeit some of their rights and freedoms and ability to influence power if the government guarantees them minimal stability and income.

There is no system of values in society which is upheld by the majority and that those in power take into consideration when making decisions; there is no constitutional consensus. Although it is clearly visible that the government is imposing and citizens are accepting, traditional for our history, etatism. The government in its search for ideas of legitimatization has begun to turn more frequently to classic national-patriotic characteristics, including perceptions of Russia's place in the world and the "Triad of Uvarov: Orthodoxy – Autarchy – Nationality", on one side and indigenous ideologies like "sovereign democracy" or "energy superpower" on the other. The result is an ideological mixture of the current "national idea", which incorporates the Russian flag, imperial crest and the Soviet national anthem; democratic rhetoric and personality cults; membership in international organizations; and aggressive actions and statements in foreign policy. Neither society nor government have answered the key question: What is Russia's past and what should its future be? Elections have the character of appointments and have become an instrument for the modification of a vertically integrated political system. It is the final rejection of competiveness and the switch to a system with an absolutely dominant party that has been legitimised through changes in election laws and political practices, weakening and repressing the opposition and transforming mass media into an effective propaganda tool.

The war with Georgia, the worsening of relations with Ukraine and Belarus and other post-Soviet countries, and minimal results in creating international political blocks have put Russia's position as a regional power and its ambitions to dominate over the post-Soviet region in question. At the same time, Russia's relations with the West have worsened. Energy shipments have become instruments of pressure and political tradeoffs.

Russia's actions have been specified by the lack of a long-term strategy, their inconsistencies with different vectors, aggressiveness, and some radically anti-Western statements. This caused reactions of the same nature from foreign partners. Other than the political elites of western countries, Russia legitimates her actions by comparing them to actions she considers unacceptable or which produce criticism, such the recognition of Kosovo or the war in Iraq. For example, the Russian justification for the movement of troops into South Ossetia is the NATO precedent in defence of Kosovo.

The continuing economic financial crisis shows just how close the economies of the world's countries are interdependent on each other, not to mention their financial markets. The crisis placed the challenge upon every country, including Russia, to create a new financial system of world economic relations. At first it seemed that the crisis should have demonstrated to our rulers that isolationism as a policy – using the logic of a "besieged fortress"[2]: the refusal to take part in international organizations, a return to the cold war, and the building of a new iron curtain – is impossible in modern conditions. In the coming time Russia will come to face a number of questions, which also require strategic vision: creating new agreements in strategic arms limitations, nuclear non-proliferation, providing international security against terrorism, becoming a member of the WTO, and signing a new partnership agreement with the EU.[3]

The war with Georgia, recognition of South Ossetia and Abkhazia as independent states, and aggressive anti-western rhetoric significantly undermined our country's credibility and made solving problems of foreign economic policy even more difficult than they already were. Militaristic public relations such as publicising tests of ageing weapons and ballistic missiles, and naval raids to the shores of South America, are meant to be domestic policy but are also a demonstration for the world. Politicians and the military passionately exaggerate the myth of a military threat and the necessity of new defensive pacts. Symbolic friendship with leaders of non-democratic countries does not gain any sympathy. All the while, the number of allies decreases.

Russia and Ukraine's "gas wars" not only became a threat to the system of energy supplies but also led to supply cut-offs in countries

2 Шевцова Л. Логика осажденной крепости. // Ведомости, 24/9/2008. Тренин Д. "Москва мускулистая": одиночество отдельно взятого центра силы. Брифинг Центра Карнеги. Том 11, выпуск 1, январь 2009.

3 Ядерное распространение: новые технологии, вооружения и договоры. Под ред. Алексея Арбатова, Владимира Дворкина. М. 2008.

of southeast Europe and furthered the fear that Russia will intensify the policy of using energy shipments as a means of political pressure under the slogan of "energy superpower". In the context of the clear anti-western direction, Russia's role in the G8, PACE, OSCE, and the Russian NATO consultative council became unclear and contradictory.

Growing distrust and even fear grips not only the political elites but also the majority of the citizens of other countries, who have become hostages of the ambitions of Russia's politicians.

This isolationist and militaristic policy contradicts modern tendencies of openness and globalization and of the interdependence of nations, as the financial crisis demonstrated. It also contradicts the country's own practical-pragmatic interests which have the difficult problem of redefining a system of treaties, related to military balance on the one hand, and dependency on the consumers of the country's energy resources, primarily oil and gas, on the other.

In the course of the last years Russian oil and gas companies have been gradually developing into international players: Gazprom, LUKOIL, RAO EЭC, Rosneft and Surgutneftegaz are expanding beyond Russia's borders. Gazprom is attempting to reach its end-consumers in Europe and expand into Asia. Rosneft produces oil in Kazakhstan and Algeria, plans to build an oil refinery in China, and will also get a stake in the Miro refinery in Germany.

Nevertheless, the internal problems of the fuel and energy companies remain unsolved.

A sharp drop in prices on energy resources exacerbated earlier symptoms of stagnation in oil production that first appeared in 2005. Natural gas production grew only by 14% from 2000 to 2008. At the same time 3 large pipelines were initiated: "Blue Stream", and the "North" and "South" streams, which will increase the export potential by 75%. The filling of these pipelines to capacity is still under question. At the same time, a reduction in the demand for gas resulted in decreased gas production and was a serious blow to Gazprom's investment abilities. Instead of investing in new project resources, the windfall of the last years, caused by advantageous conjuncture and high oil and gas prices, was spent in pursuing useless mergers and acquisitions. A number of private holders and foreign investors were ousted. Government companies, who, with the help of the authorities, captured leading positions in the oil and gas industry, took huge debts on themselves and other responsibilities. Gazprom spent only 18.5 billion dollars on the development of new fields even though"South Stream" by itself needs 24 billion dollars. In this state, the Russian oil and gas industry are now confronted by the declining demand for hydrocarbons and prices will remain moderate for the next years. Now nobody is talking about massive investments; that means no growth recovery.

The last years were distinguished by the expansion of government and other "huge" economic projects with a dominating political-ideological context and suspicious economic efficiency – or were predetermined to be ineffective economically.[4] Some of these projects were the Olympic sites in Sochi, additions to the atomic power plant on Russkij Island, the Moscow–St. Petersburg Highway, the creation of 4 "gaming zones", development of nanotechnology, and others. All of these projects were realized without perspective planning, calculating results against expenses; they thus require more and more financial investments.

It is not unusual that these projects impact the environment in a negative way (the wildlife reservations territory in Sochi was reduced in size, the Khimki forest was cut down for the construction of the Moscow–St. Petersburg highway, and so on); the country is becoming less ecologic this way. A political dead-end was created after the "five day war"; the recognition of Abkhazia and South Ossetia as states and demonstrative economic support has and will continue to have serious local, regional and global consequences.[5] Russia practically took complete responsibility on itself for the development of events in the conflict region, including the actions of the administrations of Abkhazia and South Ossetia and their military contingents. Instability in the Caucasus and Transcaucasia

4 Иноземцев В. Бесперспективные проекты. // Ведомости. 10/8/2009.
Бунин И. Военные действия в Закавказье и политическое развитие России // politcom.ru 1/10/2008.
Захаров А.Имперский федерализм.// Неприкосновенный запас, 57, 2008.
Доклад Института современного развития: Демократия: развитие российской модели. Под общей ред. Юргенса И. М., 2008.

5 Бунин И. Военные действия в Закавказье и политическое развитие России // politcom.ru 1/10/2008.

increased. Problems with unrecognized states and border disputes between them have intensified since the breakup of the Soviet Union. Disputes inside Russia concerning the borders of republics have continued, especially where there were armed conflicts in the past and in suburban regions. In a time of decreasing income, economic support of South Ossetia does not seem to be without problems.

There is no hope that the legal independence of Abkhazia and South Ossetia will become a fact of international law; Russia was not supported by its few allied countries. Russia's actions and statements were perceived as a threat to return to the times of the cold war by the international community. Our country's membership and role in international organizations and already made international agreements were put in question, complicating relations with a number of countries.

Creating a system of international safety guarantees conditions for a normal life and the return of refugees for the citizens of Georgia, South Ossetia, Abkhazia; after more than half a year, it is still being examined. The international community still has not answered the question as to who was responsible for the war and what the repercussions will be for the involved sides. The goal of establishing peace in the caucuses is still dependent on the interests of different countries and blocks; thousands of people remain hostages of political games which have ended in tragedy more than once.

The war with Georgia became a first and very demonstrative precedent of Russia's neo-imperial politics in relation to former Soviet republics. "The shadow of the Russian Federation, ever more powerful and authoritarian, has become an instrument of self-assertion for the CIS area in our time, as the symbols of the Russian Empire and the Soviet Union were before in relations with Europe", notes Andrei Zakharov.[6] Instead of employing various historically established business, cultural, political relational links to build mutually effective relations, Russia, laying claims to dominate the post-Soviet region, is using not only political and economic pressure now, but also military force, pushing away Russia's natural allies and destroying the potential accumulated through the centuries.

6 Захаров А.Имперский федерализм.// Неприкосновенный запас, 57, 2008.

The aftermath of the financial crisis, the military operation in Georgia, and the recognition of Abkhazia and South Ossetia clearly showed that a "revenue state" whose economy is based exclusively on high prices for energy resources cannot be stable, while foreign policy and changes in the world economy are tightly related to the events in our country.

The paradox is in the fact that Russia's ruling class and groups close to them are objectively interested in "becoming part" of international structures and affirming their positions, including lucrative contracts that attract (especially Western) investments and increase the possibilities for contacts in education and culture.

However the opposite development, which is conditional upon the dependence of Russia's foreign policy on the characteristics of regime development, is presently occurring.

The Russian authorities, facing the decision between modernization and mobilization – in other words between liberalization of the regime with the return of democratic institutions and norms or tightening the regime – are acting in different ways. On one side, Medvedev as president frequently uses liberal rhetoric, which causes optimism in the liberal camp; on the other, mobilization mechanisms are employed.

Mobilization presumes a chain of events increasing the sphere of their radicalization: military conflict – military action – war; declaration of exterior enemies – militarization – isolation – search for the interior enemy that helps the exterior enemy – repressions. In other words the logic of expanding an authoritarian regime towards tougher authoritarianism with elements of totalitarianism presupposes the final rejection of democratic institutions and maximum control of the government over all areas of life.

The possibilities of solving the task of mobilization, which is connected with the final suppression of the opposition and the permanent search for interior enemies, have significant limitations.[7] The effects of propaganda have their limits, because one needs to show not only militaristic declarations but also negotiations; this changes the simple worldview. The limitations of the mobilization scenario

7 Доклад Института современного развития: Демократия: развитие российской модели. Под общей ред. Юргенса И. М., 2008.

are also connected with the technical unpreparedness of the elites to implement it. Under the conditions of "manual management", internal conflicts on the lower levels of the ruling group and the lack of a non-corrupt repression apparatus broad repressions are not realizable. All the while a simple non-contradicting ideology is needed from the authorities to legitimate such actions. Eurasianism in the style of Dugin or Prohanov does not fit this role. Communism and socialism in this role is already impossible in Russia. The authorities will be afraid to use Russian fascism because it is very dangerous in a multiethnic country where a part of the elite are non-Russians. A significant part of (active) society has personal experience with the West and values those possibilities, beginning with tourism and personal and educational trips ending with the possibility of having a business. Any attempt to return to the elements of the iron curtain can incite a negative reaction in these groups.

So, under current conditions it is very difficult to realize the mobilization scenario "in its pure form". The scenario being carried out now is using traditional, in form democratic institutions and democratic rhetoric for legitimating authority while emasculating them and maximizing control at the same time, with elements of mobilization.

Limitations for tightening the regime further exist internally and externally. External limitations are necessary in order to "appear proper" in the international community: the necessity the existing elections, political parties, NGO's, critical mass media.

Internal limitations are connected with the composite style of the country, the presence of vertical and horizontal conflicts and civilization fault lines, lines of social tension that can resonate at any moment. To this, latent conflicts within the elite are added.[8] The domination of informal agreements in decision-making at all levels and lack of fixated rules of the game makes even medium-range planning almost impossible. There is only one restraining factor in all lines of conflict: Putin (plus Medvedev) legitimizing

8 Захаров А. Имперский федерализм. // Неприкосновенный запас, 3, 2008, pp. 65–83; Тишков В.А.Забыть о нации (Постнационалистическое понимание национализма) // Вопросы философии, 9, 1998 (цит по: www.portal.rsu.ru/culture/rostov.doc).

their actions through their personal support. The "signals" coming from the top are understood and interpreted in various ways by the local bureaucracy. The results may have characteristics opposite to those expected.

The logic of regime development pushes to close to totalitarian attempts to control the mass media including the internet, and driving popular journalists out of the significant mass media. Attempts to control the internet are taken up every so often. However, complete control over mass media in today's information society and the special functioning's of the internet is an unsolvable task.

The main effort is directed toward the work of the propaganda machine of the main TV channels. At the same time, out of the presumption of the critical mass media's insignificance due to the small size of their audience, small opposition media that have limited broadcasting range or circulation ("Эхо Москвы", "Новая газета" and so on) and opposition internet resources are allowed. The internet is outside the area of control, but using this source requires a certain level of culture, even a literate user is not always able to distinguish between facts and their interpretations, truth and lies. The government is constantly expending substantial effort to create imitations of free and critical resources.

The problem of the propaganda machine's effectiveness of impact on the audience is connected with the inertia of the machine that "overworks"; this damages the government when short-range goals are changed because a schizophrenic picture of reality is created. At the present, an overshoot of heavy-handed and aggressive anti-Western positions dominates. This forces the thinking audience to search for alternative sources of information that are located in the "zone of trustworthiness".

After passing law on NGO's and creating the Public Chamber, the significance of this sector of society decreased while controllability increased. Traditional human rights groups, whose number has decreased, are not an exclusive source of information about Russia in the west and are in no condition to become a foundation for the creation of a meaningful opposition (create conditions for a "colour revolution"). That is why, despite the fears and preventive measures of the government, they do not pose any threat.

New epicentres of citizen's activity are related to attempts of (non-formalized) association to solve problems essential certain groups (initiative groups, auto owners, protesters and others). These groups put pressure on the government and are minimally structured, if there is a leader. They distrust formally existing organizations of civil society, but if they see specific results from cooperation they are ready to establish contact. They are the real reserve for the organization of a civil society, including democratic parties. Their existence is a necessary condition to liberalize the political policy.[9]

The main danger for the regime is the decrease of controllability and the uncontrolled development of events that relate to other processes; all the while, the above mentioned institutions can be regulators of these processes to a limited degree.

These processes are:

- The multidimensional results of the financial and economic crisis: a decrease of GDP growth, the deterioration of the investment climate, a decrease in incomes, the downsizing of workplaces, and so on. These have different impacts in different industries and regions. In any case, this only increases the gap between the income and possibilities of different social and regional groups, in different types of settlements. The escalation of tension in the Northern Caucuses is not declining due the practical insolvability of the problem of political stabilization and economic development in the conditions of a clan system, inter ethnic conflicts, and consequences of historical wounds (problems with the aftermath of military actions, borders, refugees, repressed peoples, etc.).
- Ineffectiveness of the system of administration, a high level of corruption in government institutions, the lack of a clear long-term strategy, using the imitation of national projects and programs in this field, that have little connection to reality. The statement of "nanotechnology" as a goal has nothing in common with the problems of the majority of the nation; the lack of roads, home improvement, a working health system, pensions, etc.

9 Ворожейкина Т. Надобно годить. // Новая газета, 5, 2009, 5, 21/1/2009.

- Intrusion of the state in economic processes, pressure on business, the level of trust and stimulus for development in private companies.
- Ineffectiveness in the conditions of a growing gap in the level of economic development in different regions and different types of settlements; unequal living standards of the population, a non-unitary system of administration, the existence of latent conflicts between the center and regions; all can escalate at any moment to the point of reviving dicussion of the benefits of separatism for the regions' elites.
- The vulnerability of property rights and economic raiding is the protest potential of owners of small and medium sized businesses (in connection with packed-in construction, the growth of utility payments etc.)
- Over-exploitation, vulnerability of labour rights, increased activity of independent trade unions, increased frequency of protests, lack of institutional mechanisms for resolving conflicts between workers and employers.
- And finally, the price increase on food products and fuel.

All of this against the background of the necessity of solving the problems of modernization, transition to an economy that is not resource based; these are not only evident to the majority of the population but have also been declared to be primary goals by the country's management.

The reaction of Russian authorities can be classified as a waiting policy. Holes were mended while waiting for radical improvement in the West and, first of all, an increase in the prices of energy resources. The measure taken – giving money to the banks first, and only afterwards supporting the real economic sector – had no effect.

Russia's ascendance into the world financial crisis inspired the idea in several experts and public figures that the "quiet agreement" that regulates relations between the government and society, when society gives up freedoms in exchange for part of the "oil revenue", in other words personal income growth, needs to be renegotiated. The initiators of the discussion were based on the presumption that the current contract is socio-economical in its nature. It was stated that Russia's citizens gave up their rights in exchange for the increase of personal wealth. However since the crisis began, oil revenue has decreased and income is declining, therefore it is necessary to term the contract as void and citizens should have their rights and freedoms returned to them. After the limitations of the government's financial possibilities became visible, the need to modernize and develop competitiveness not only in the economy but also in politics, its "defrosting" became evident to a broad circle of experts, the interested part of society and representatives of the government. In connection with the crisis, the system needs to be (and will have to be, to put it mildly) changed. To be honest, a problem arises; at first glance no intentions to "deep freeze" the system can be seen. The readiness for change has been inserted into the "system" from the very beginning. To elaborate, changes began long before government and society felt the impact of the financial crisis from the moment there was rotation in power, in fact, even a bit earlier. The new president and the new construction of higher authorities is the first step. The second is amending the constitution, and the third Medvedev's pack of political reforms. The main specific of the ongoing political reform is its conservatism: it stretches the change through time, assuming that society will get the possibility to get used to them, and transformational institutions become rooted.

Anticipation of mass protests as a result of the social impacts of the crisis, which could have sped up these reforms, were not justified. In any case, until the summer of 2009, in spite of the downsizing of jobs, decreasing production and working time, that affected a number of regions as well as mono-cities, protests were local and never became massive. The crisis affected different region and types of settlements in various ways. If its effects were felt in developed industrial areas, in distant regions of the periphery – small regional centers with inadequately functioning economic infrastructures – the crisis was experienced as an increase in inflation.

The super centralization that characterizes the relations between the center and the regions led to the "Pikalevo effect", in which the mobilization of government resources, including the personal attendance of the prime minister, is necessary in order to solve a local problem of unpaid salaries. Changes in international policy, the system of international relations, multiplied by the effects of the crisis resonates on all of these problems, making the existing internal problems worse.

It is unlikely to expect that other significant international players, be they the USA, EU, NATO or other international organizations in the visible future can directly influence the Russian economy or political system, as in the beginning of the 1990s during the time of the Washington consensus and large loans for Russia.

The fashion of *Realpolitik* and the temptations of isolationism as a result of the crisis do not cancel the goals of arms limitation, the fight against international terrorism, and the need for efficient international organizations. Russia will continue to play an important role in this process. That is why the questions of the basic values of the past and the future, and in what direction the country should move, have to be answered by the elite and the active parts of society.

* * *

The second half of Vladimir Putin's decade, which began in 2006–2007, was characterized by a increasing feeling of disappointment and aggravation and doubt in the possibility to agree on anything. A policy of increasing potential independence was taken in order to be prepared for any twists and turns of the international system. As a result Russia increased her weight in the international arena and forced others to consider her more seriously, while at the same time the potential of external countermeasures and alienation increased.

This is substance of the Putin paradox. The goals that he put forth before himself and the country from his first months of presidency have not been fulfilled by the second term – this is partially his fault, because he perceived the rules of international communication in a specific way, but mostly they were not reached because of an unfortunate flow of external events. At least he excelled in what he was not aiming for and what became a reaction to the top orders of business. Here also, the merits can be divided. The personality of the head of state played its role here, but a lot of it is due to the sudden events which have been unfolding in the international arena since the beginning of the 21st century.

Russia's foreign policy during Vladimir Putin is very personalized, especially in the latter period. The personal relations accumulated through the years of presidency with various leaders and countries frequently exceeded the necessary level for a pragmatic government approach. Personal traits and characteristics set the mood of relations with the huge state as a whole. The deepening feeling of Russia's unpredictability is related to this. The leader's emotionality is more evident than the conceptual base of a foreign policy approach - which is adequately stable and sequential.

The principle question consists in what comes first - Putin's policy, causing a corresponding reaction from the outside, or the surrounding situation, provoking the Russian leadership to take certain actions? Perhaps in the future, serious-minded commentators will be able to answer that question, abstracting from the personality of the main actor. Whatever it may be, Putin doesn't seem to be very present on the international stage at the beginning of the 21st century.

He is notably outside ideology but also a symbol of the decreasing significance of ideology in world politics in favour of classic relations between states. Vladimir Putin is blamed for bringing back the spirit of the political rivalries of the past. However, his main difference is that he is untraditionally frank in his statements and straightforward in his means, and not in the fact that others are not attracted to tough competition. And finally Putin is blamed for overplaying the meaning of external effects and symbolic actions. This is not really worth a comment against the background of the public relations and glamour technology that has infested the politics of the largest states.

Galina Michaleva and Andrey Ryabov

The Authors

Dmitrij Furman is political scientist, head of the Centre of the CIS in the Institute of Europe at the Russian Academy of Sciences. He is the author of books about political processes in Azerbaijan, Baltic states, Belorussia and other countries of the post-Soviet space, and he is the author of publications on other different political topics, his articles are published in the "Russian Magazine", in the "Nezavisimaya gazeta" newspaper, "Free Idea" magazine.

Maria Gordeyeva is professor at the History, Political Science and Law Department in the Russian State University for the Humanities. She also works there as the Vice Dean of the History, Political Science and Law Department. Her research interests are: conflict management: the prevention, management and transformation of international conflicts; negotiations: techniques and approaches; identity studies: symbols of identity and their impact on political behaviour as well as communication and political action. She is the author of more than 20 works.

Fyodr Lukyanov is journalist and political scientist. Since 2002 he has been the editor-in-chief of the "Russia in Global Affairs"-magazine about latest global developments and Russia's role in them. He also writes articles for different newspapers and magazines, including The Moscow Times, United Press International, Guardian and the website "Gazeta.ru".

Galina Michaleva is the executive secretary of Political committee of the party "Yabloko", chairwoman of the organization of women in "Yabloko" and professor at the History, Political Science and Law Department at the Russian State University for the Humanities. She is author of more than 100 books and scientific publications in Russian English and German, including: „Europa und Russland. Annäherungen und Differenzen" (2007), „Politische Parteien in Russland.

Dokumente und Kommentare" (2003), „Politik und Kultur in der russischen Provinz" (with S. Ryshenkov and others) (1999), „Parteien in der russischen Provinz" (1998).

Vladimir Milov is a Russian independent politician, economist, and energy policy expert, member of the governing board of the Russian democratic opposition coalition "Solidarity", president of the Institute of Energy Policy, a Moscow-based independent consultancy, and former Deputy Energy Minister of the Russian Federation.

Alexej Yablokov is a prominent Russian environmentalist, former member of USSR parliament and environmental advisor to the Russian President Yeltsin and to the Gorbachev administration. Alexej Yablokov has been active in many efforts to reveal conservation and pollution challenges in Russia addressing such issues as illegal whaling. He focuses on the problems of radiation contamination, particularly in marine ecosystems and biology of the marine mammals. In 2004 he became the chairman of "Green Russia" party and the consultant of the Russian Academy of Sciences. In June 2008 he was elected as a member of the Political Committee of the "Yabloko" Party.

Grigory Yavlinsky is a Russian politician and economist. He is well-known as the author of the "500 Days" project – a programme for the transition of USSR economy into free-market type of economy. To implement the program, Grigory Yavlinsky was appointed Deputy Chairman of the Council of Ministers of the RSFSR and Chairman of the State Commission of the USSR for Economic Reform. In 1993 he began working in the state Duma; he was the leader of the successful first, second and third convocations of the Duma. Grigory Yavlinsky created Russian Civil Democratic Party "Yabloko" in 1995 and until year 2008 he was the party leader. He ran twice for president – in 1996 he came in fourth place and in 2000 he was third. He is the author of over 60 books and publications translated into English, German, French and Japanese, including: "Analysis of the Economy of the USSR" (1982), "The Grand Bargain" (1991), "Nizhni Novgorod Prologue. Economic and Politics in Russia" (1992), "Lessons of the Economic Reform" (1994), "Russia's Economy: The Heritage and

Possibilities" (1995), "Russia's Phoney Capitalism" (1998), "Incentives and Institutions".

Natalia Zubarevich is professor at the Geographic Department of Moscow State University. She participates as the head and chief executive in programs of the Ministry of Economic Development and Trade of the Russian Federation, the Ministry of Labour and Social Protection, and also in the international projects and programs. She is the head of the regional programs at the Independent institute of Social Policy, and the expert of the Program of development of the United Nations and the Moscow representation of the International organisation of work. Natalia Zubarevich wrote about 140 scientific publications as the author or co-author of them, including the "National Human Development Report 2006/2007 for the Russian Federation", "Index of Development of Human Potential of Russia's regions" (1998), "Social Development of the Regions of Russia: Problems and Tendencies of the Transition Period" (2003) and others.

Alexei Zudin is political scientist, the senior lecturer of the chair of public policy in the State University - Higher School of Economics and the head of the department of political programmes of the Centre for Political Technologies. He is the author of the following works: "The Russian Presidential Election in 2000 and its Implications" (2000), "Oligarchy as a Political Problem of Russian Postcommunism"(2000), "'Putin's coalition': Potential and Internal Restrictions"(2000), "Union of Right Forces and Elections in the State Duma: Political Frameworks of an Electoral Victory" (2000) and other scientific publications.

Nina Poussenkova is scholar-in-residence of the Carnegie Moscow Center. She cooperates with WWF-Russia and is a senior policy advisor of its Trade and Investment Program. Dr. Poussenkova is a member of the advisory board of the Greening of Industry Network. She has been involved in the Russian oil industry since 1991. She is among the authors of the four-volume research report Russian Oil (Salomon Brothers, 1996) - one of the first major publications about the Russian oil sector.

Andrey Ryabov is scholar-in-residence of the Carnegie Moscow Center, a deputy director of the Center for Political Science Programsat Gorbachev Foundation, political observer of the newspaper Vek, deputy editor-in-chief of the journal Vestnik of Moscow University (political sciences series), and member of the Russian Association of Political Science.

INTERDISCIPLINARY STUDIES ON CENTRAL AND EASTERN EUROPE

Edited by
Christian Giordano, Nicolas Hayoz & Jens Herlth

This new series focuses on the political, economic and cultural changes in Eastern, East-Central and South-Eastern Europe. It publishes scholarly papers presented at the "Eastern European Day" conferences held at the University of Fribourg. These annual conferences examine the various East European countries. Moreover, the series also presents a forum for other conferences held in Switzerland on topics related to Eastern Europe. It aims to reflect on what could be considered the rapprochement of the different branches of Swiss research on Eastern Europe (history, political science, philosophy, Slavic languages). The articles are written in English.

Band 1 Nicolas Hayoz, Andrej N. Lushnycky (eds)
Ukraine at a Crossroads.
269 pp. 2005.
ISBN 3-03910-468-3, US-ISBN 0-8204-7157-7

Band 2 Nicolas Hayoz, Leszek Jesień, Wim van Meurs (eds)
Enlarged EU – Enlarged Neighbourhood.
Perspectives of the European Neighbourhood Policy.
392 pp. 2005.
ISBN 3-03910-624-4, US-ISBN 0-8204-7167-4

Band 3 Nicolas Hayoz, Simon Hug (eds)
Tax Evasion, Trust, and State Capacities.
How good is Tax Moral in Central and Eastern Europe.
379 pp. 2007.
ISBN 978-3-03910-651-6, US-ISBN 978-0-8204-7168-6

Band 4 Andrej N. Lushnycky, Mykola Riabchuk (eds)
Ukraine on its Meandering Path Between East and West.
Interdisciplinary Studies on Central and Eastern Europe.
210 pp. 2009.
ISBN 978-3-03911-607-2

Band 5 Leonidas Donskis (ed.)
A Litmus Test Case of Modernity.
Examining Modern Sensibilities and the Public Domain in the Baltic States at the Turn of the Century.
314 pp. 2009.
ISBN 978-3-0343-0335-4

Band 6 Marta Dyczok, Oxana Gaman-Golutvina (eds)
Media, Democracy and Freedom.
The Post-Communist Experience.
246 pp. 2009.
ISBN 978-3-0343-0311-8

Band 7 Josette Baer (ed.)
Revolutions, Business and the State.
Recent Political and Economic Developments in Eastern Europe.
Forthcoming.
ISBN 978-3-03911-773-4

Band 8 Galina Michaleva, Andrey Ryabov (eds)
Russian Challenges.
Between Freedom and Energy.
240 pp. 2011.
ISBN 978-3-0343-0529-7

FSC
www.fsc.org
MIX
Papier | Fördert gute Waldnutzung
FSC® C083411

Kolibri 360 GmbH
Ferdinand-Jühlke-Straße 7
99095 Erfurt, Deutschland
produktsicherheit@kolibri360.de